Conservation of Plant Genes

DNA Banking and *in vitro* Biotechnology

Conservation of Plant Genes

DNA Banking and *in vitro* Biotechnology

Edited by

Robert P. Adams

Janice E. Adams

Plant Biotechnology Center
Baylor University
Waco, Texas

ACADEMIC PRESS, INC.

Harcourt Brace Jovanovich, Publishers

San Diego   New York   Boston   London   Sydney   Tokyo   Toronto

Academic Press Rapid Manuscript Reproduction

Academic Press, Inc.
San Diego, California 92101

United Kingdom Edition published by
Academic Press Limited
24–28 Oval Road, London NW1 7DX

Library of Congress Cataloging-in-Publication Data

Conservation of plant genes : DNA banking and in vitro biotechnology /
    edited by Robert P. Adams, Janice E. Adams.
        p.   cm.
    Papers from the organizational meeting of DNA-Bank Net held in the
    Royal Botanic Gardens, Kew, London on April16-17, 1991, including
    those prepared for the meeting but whose authors could not be
    present.
    ISBN 0-12-044140-3
    1. Germplasm resources, Plant--Congresses.   2. Gene banks. Plant-
-Congresses.   3. Plant biotechnology--Congresses.   I. Adams, Robert
P. II. Adams, Janice E.   III. DNA-Bank Net. Meeting (1991 : Royal
Botanic Gardens, Kew)   IV. Title: DNA banking and in vitro
biotechnology.
QK981.7.C66   1991
631.5'23--dc20   91-30629
                                                                     CIP

PRINTED IN THE UNITED STATES OF AMERICA
91 92 93 94   9 8 7 6 5 4 3 2 1

CONTENTS

DNA Bank-Net Workshop Participants
Royal Botanic Gardens-Kew   April 1991

Dr. Robert P. Adams, Plant Biotechnology Center, BU Box 97372
 Baylor University, Waco, TX 76798 USA
      Phone 817-755-1159, FAX 817-752-5332

Dr. Daniel K. Abbiw, Botany Department, Box 55, University of
 Ghana, Legon, Ghana, West Africa

Dr. Luiz Antonio Barreto de Castro, CENARGEN/EMBRAPA, Parque
 Rural, CP 102372, Ave. W5 Norte, W 70770, Brasilia DF,
 Brazil

Dr. Mike Bennett, Keeper, Jodrell Labs, Royal Botanic Garden,
 KEW, Richmond, Surrey TW9 3AB, England

Mr. Tony Cox, Jodrell Labs, Royal Botanic Garden, KEW,
 Richmond, Surrey TW9 3AB, England

Mr. Abebe Demissie, Plant Genetic Resources Center, P.O. Box
 30726, Addis Ababa, Ethiopia

Dr. David Giannasi, Dept. of Botany, University of Georgia,
 Athens, GA 30602

Dr. Toby Hodgkin, Research Officer, Genetic Diversity IBPGR,
 c/o FAO of the UN, Via delle Sette Chiese 142
 00145 Rome, Italy

Prof. Lin Zhong-ping, Division of Plant Molecular Biology
 Institute of Botany, Academia Sinica, Beijing 100044,
 China

Dr. James Miller, Missouri Botanical Garden 2315 Tower Grove
 Ave., St. Louis, MO 63166

Dr. Simon Owens, Jodrell Labs, Royal Botanic Garden, KEW,
 Richmond, Surrey TW9 3AB, England

Dr. Bart Panis, Laboratory of Tropical Crop Husbandry, Catholic University of Leuven, Kardinaal Mercierlaan 92. B-3001 Heverlee, Belgium

Dr. Steve Price, Industrial Liaison Officer, Office of Intellectual Property, Iowa State University, Ames, Iowa, 50010

Dr. Phillip Stanwood, National Seed Storage Lab, USDA, Colorado State University, Ft. Collins, CO 80523

Dr. Dennis Stevenson, The New York Botanical Garden, Bronx, New York 10458-5126

Dr. Peter Strelchenko, N. I. Vavilov Institute of Plant Industry, 42, Herzen Street, 190000, Leningrad, USSR

Dr. Victor M. Villalobos, Director, Programa Mejoramiento Cultivos Tropicales, CATIE, Turrialba 7170, Costa Rica

Mr. Zheng Sijun, Dept. of Agronomy, Zhejiang Agricultural University, Hangzhou, 310029, Zhejiang, China

PREFACE

How does one begin to grasp the extinction of a species? For most scientists, extinction is an academically interesting subject but of no personal grief as in the case of seeing an old friend die. No doubt, for me, I had maintained an academic interest in rare and endangered plants in concert with the 'environmental' awareness that biologists wear. Yet, it was not until about 1987 that I realized that the life of a friend of mine, *Juniperus barbadensis* L., was very endangered in St. Lucia, BWI. The following summer, I received a letter from a colleague in St. Lucia that an illegal fire on the top of Petit Piton had probably destroyed the last known population of *Juniperus barbadensis* in the world (the species was cut out on Barbados Island before 1700!). That was when extinction became real and personal. Faced with new opportunities to examine DNA from *Juniperus*, I was immediately struck by the fact that I had collected fresh leaves of *Juniperus barbadensis*, transported them to my laboratory, made herbarium specimens, extracted the volatile oil, but failed to keep any frozen leaves for future work. One can seldom see the future.

So began my quest four years ago to attempt to get the world's great botanical institutions interested in collecting and storing plant materials for subsequent DNA extraction and utilization. DNA Bank-Net is the result of the past four years' conversations, letters, proposals, FAXes, and seminars.

The organizational meeting of DNA Bank-Net was held at the Royal Botanic Gardens, Kew, London on April 16-18, 1991. The meeting was attended by eighteen invited scientists who represented a range of institutions and interests from strictly academic to very applied. As can be seen from the group reports, this resulted in rather spirited discussions. Nevertheless, it was felt that this diversity was necessary in order that DNA Bank-Net not be too narrowly focused.

This book contains papers that were presented at Kew or for a few invited scientists who could not attend, we are also publishing their papers. The papers were sent to me on floppy disks and then converted to the Volkswriter Word Processor

format (Volkswriter, Inc.) using Word for Word (DS Technologies, Inc.). The papers were edited for spelling and grammar
but left essentially as written. The ideas, opinions and
phraseology are that of the authors not the editors.

The papers range from applied to somewhat academic, from
conservation *per se* to *in vitro* cryopreservation to DNA storage methods and PCR amplification of fossil DNA. Our goal is
that this book will act as a catalyst to encourage the immediate procurement and storage of DNA from all kinds of organisms, even as work continues to preserve species *in situ*. We
now have the opportunity to purchase some 'genetic fire insurance'. The technology is sufficient; all we need is the
vision.

It is a pleasure to acknowledge Helen Jones (Helen Jones
Foundation) for her vision of the world and the future. The
Jones Foundation has been a consistent supporter of our work
on DNA preservation at Baylor and without that support, the
past four years of research leading to the meeting would not
have been possible. In addition, the Conservation, Food and
Health Foundation and the Wallace Genetic Foundation directly
supported the meeting and their support of the participants'
expenses is greatly appreciated.

Special thanks to the Royal Botanic Gardens, Kew, Prof.
Mike Bennett, Director, Jodrell Labs, Prof. Ghillean Prance,
Director, Kew for hosting the conference. We are also
indebted to Dr. Simon Owens and Tony Cox for their assistance
during the meeting.

EFFORTS TO CONSERVE TROPICAL PLANTS - A GLOBAL PERSPECTIVE

Vernon H. Heywood

Plant Conservation, IUCN - The World Conservation Union
199 Kew Road
Richmond, Surrey, TW9 3BW United Kingdom

Summary - An overview of man's current efforts to preserve plant diversity, both through *in situ* and *ex situ* conservation, is presented. Several approaches are reviewed,
such as the IUCN Biodiversity Conservation Strategy, IUCN Protected Areas Program, Botanic Gardens Conservation Secretariat, the Centers for Plant Diversity and the Rapid Assessment
Program.

INTRODUCTION

A great amount of attention is focused today on the
threats to biological diversity, especially in the tropics and
in particular on the plight of the rain forests which are currently estimated to be disappearing at the rate of 10.5 million ha per year for closed forest and 20.4 million ha annually for total forest loss, according to the latest estimates
of WRI, supported by FAO and other sources. The loss of forest is interpreted here in the sense of conversion of forested
land to permanently cleared land for agriculture, pasture,
industrial development or other uses, or to a cycle of shifting cultivation. The tropical forests are believed by Wilson
(1989) to contain about half of the world's species in an area
representing only 7% of the world's land surface.

It should be noted, however, that other tropical ecosystems are also suffering severe depredation, such as savannas,
scrubland and some mountain communities. It is inevitable
that this massive loss of habitat will lead to local or total
extinctions of many species, although just how many is a
matter of debate. Estimates by biologists such as Raven
(1987, 1988), Wilson (1988), Myers (1988a,b), Simberloff
(1986), suggest that tens of thousands of species face imminent extinction during the coming decades, although such pre-

dictions are to be regarded as hypotheses and are based on extrapolations from indirect evidence. What is certain is that tens of thousands of species are suffering loss of genetic variation (genetic erosion) of their populations.

In the light of this situation it is incumbent on us not just to issue cries of alarm but to implement programs of action. This paper will review the various ways in which conservation of tropical plants is being undertaken. It is important to stress that an integrated program of conservation should be undertaken in which whatever technique or action that is necessary to the situation is applied, whether in situ or ex situ. As Falk (1990) notes, the full spectrum of conservation resources must be employed in a coordinated manner without relying exclusively on any single approach.

COMPLETING THE INVENTORY

The basis for all conservation action is sufficient knowledge of the diversity of the organisms concerned and the ecosystems in which they occur. For this reason we need to complete the inventory of plants (and animals and microorganisms) which is of course the primary function of the taxonomist.

As many commentators have pointed out in recent years, the inventory is far from complete and although we cannot give precise figures, the number of organisms that have described scientifically so far by taxonomists is of the order of 1.5 million, but this represents 5-15% of the total number of species that are believed to exist at the present time. There are several reasons for the considerable uncertainty as to the total number of species extant today – the efforts of plant taxonomists have tended to focus on particular groups such as the higher plants (flowering plants and ferns) while nonvascular plants such as bryophytes, algae and fungi have received less attention. Large numbers of soil fungi (probably hundreds of thousands of species) remain to be described, while in the animal kingdom, the number of undescribed species probably runs into millions.

The flowering plants are well known in the sense that perhaps only 5% remain to be described scientifically which is not surprising in view of their key role in the structure of

many ecosystems and the benefits many thousands of them provide to humankind. It has to be noted, however, that for the
majority of the approximately 250,000 that have been
described, only the basic facts of their morphology and distribution is known. Little effort has been made to study the
reproductive biology, cytology, genetics, population structure
or chemistry of most tropical species. Their detailed distribution is unknown and consequently it is often difficult to
ascertain their conservation status.

The majority of flowering plants and ferns are tropical in
distribution - about 90,000 in Latin America, 60,000 in Asia
and 40,000 in Africa (including Madagascar). For many tropical countries no complete Flora exists nor even local Floras
or identification manuals. Some Flora-writing projects are
reappraising their timetable and scope in the light of slow
progress being made towards completion. An example is the
Flora Malesiana which has failed to publish more than 20% the
Flora in 40 years and the realization by the editors that it
would take a total of 200 years to cover the 25,000 species
estimated to occur in the region. There is typically a shortage of facilities, such as herbaria, libraries and botanic
gardens and of trained staff in most tropical countries and
generally an inverse relationship between floristic diversity
and institutional resources.

Several innovative attempts are being made to remedy the
situation described above. A number of rapid inventory techniques have been devised that will allow information to be
gathered in the field and processed in the most effective manner possible. The use of computerized databases and word processing in the production of inventories and Floras is nowa-
days commonplace although there are still problems with data
standards and information transfer to be overcome. Such
matters are being tackled by the International Taxonomic Database Working Group for the Plant Sciences (TDWG). A more
recent development has been the Q taxa project pioneered by
Professor Arturo Gomez Pompa and his team at the University of
California, Riverside and in Mexico which involves the use of
CD ROM videodisc technology to store large amounts of the
information necessary for taxonomic work, including text,

illustrations, drawings, photographs of type specimens, habitat views, reprints, extracts from keys, floras and other relevant literature, all indexed and searchable.

Another novel approach has been developed by Professor Rodrigo Gamez at the Instituto Nacional de Biodiversidad in Costa Rica. This is based on the training of local, relatively unskilled local people as parataxonomists whose role is to undertake inventory work, collect and process herbarium material and gradually develop taxonomic skills under the guidance of professional staff, including expatriates on secondment for periods. Other countries have developed short-term practical training courses such as the recently inaugurated UNEP/IUCN sponsored course in taxonomic competence at the University of Reading, U.K.

Of course, inventory alone will not conserve plants or save them from extinction and taxonomists should work in close association with conservationists in ensuring whenever possible the *in situ* or *ex situ* preservation of the species and populations that are most at risk. Large herbarium collections of specimens of species that have become extinct will be of largely academic value unless techniques are devised to allow genetic material to be extracted from them that will allow their propagation. Inventory must be accompanied by conservation action as considered in the next section.

APPROACHES TO CONSERVATION

The main techniques that have been developed to conserve biological diversity have tended to concentrate on the establishment of protected areas (*in situ* conservation) and genebanks of various sorts, including some botanic gardens (*ex situ* conservation). The first approach aims to preserve habitats or landscapes, and in a general way the species that they contain, while the second approach concentrates on particular species or the genetic variation that they contain. Both have their advantages and their limitations and the recommended philosophy today is to adopt an integrated strategy for conservation as advocated by IUCN, WRI and UNEP in their Biodiversity Conservation Strategy and Programme (see also McNeely *et al.*, 1989 and Falk 1990). Such an approach addresses all

levels of biological organization, nor just species or community diversity and applies whatever techniques from the whole
spectrum of those that are available to the problem in hand
without relying exclusively on a single approach. This
involves botanic gardens, arboreta and seed banks, alongside
protected area management and habitat restoration.

PROTECTED AREAS

Legally protected areas are arranged by IUCN in eight
categories in ascending order of degree of direct human use
permitted in the area. Category I is a scientific
reserve/strict nature reserve, category II a National Park in
which commercial extractive uses are not permitted, all the
way through to category VIII which is a multiple use management area/managed area reserve which provides for the sustained production of water, timber, wildlife, pasture and out-
door recreation with the conservation of nature primarily
orientated towards the support of economic activities. Most
national governments now accept the importance of legally protected areas as part of their overall policy of land use.

The world's protected areas currently cover about 5% of
the land area, according to the data gathered by the Protected
Areas Data Unit of the World Conservation Monitoring Centre.
There are about 450 such areas in the Afrotropical realm
totalling some 86 million hectares, 676 in the Indomalayan
realm (32 million ha), and 458 in the Neotropical realm (77
million ha). These figures refer to protected areas over 1000
hectares in size. In terms of conservation of biological diversity in general and tropical plants in particular, the
effectiveness of these different kinds of protected area
varies enormously. Many of them are managed to conserve the
maximum amount of biodiversity in the area or for the conservation of particular species of animal wildlife. Few of them
are specifically managed with a view to conserving particular
plant species. The majority are unfocused at the species
level and are of limited value for the conservation of plant
genetic resources without the introduction of a targeted management regime. It is largely for this reason that we advocate the partnership between protected areas and small func-

tional botanic gardens in the neighborhood of these areas.

In addition there are many smaller areas, which are not
recorded by the WCMC Protected Area Data Unit, that play an
important part in the preservation of individual species. A
good example is on the island of Mauritius where an IUCN-WWF
project that has been running for several years under the
direction of Wendy Strahm involves establishing a network of
mini-reserves which will cover a large percentage of the
endangered plant species of the island, many of which are
reduced to small populations which are no longer able to sustain themselves by reproduction. The reserves are fenced and
weeded and *ex situ* stocks of the most endangered species are
grown in nurseries at Curepipe for restocking the natural populations. Evidence is accumulating that many species are able
to survive for periods of up to 1000 years in small populations in vegetation fragments provided that their boundaries
can be secured.

Of course the number of reserves in the tropics, as elsewhere, is quite inadequate to secure the survival of as broad
and representative a selection of ecosystems and species that
would be desirable and a doubling or even tripling of the present area under protection would be highly recommendable. It
is widely accepted today that most conservation will in future
be carried out outside protected areas and it is suggested by
McNeely *et al.* (1989) that strictly protected areas are
unlikely ever to cover more than 4 percent of the globe. They
also comment that in coming decades developments will be in
establishment and improved management of those categories of
protected areas where some human use will be tolerated or even
in the establishment of new types of protected area in
degraded landscapes which have been restored to productive use
for conservation by habitat restoration or rehabilitation.
Many reserves are inadequately protected or managed or are
subject to invasion, clearing and other kinds of illegal activities, and are thus reserves in name only. Even legal protection does not guarantee their position in some cases. A
critical aspect is the need to develop a good relationship
between the protected area and the local community.

Inventories have not been completed for most protected

areas and it is not known exactly which rare and endangered occur in them. Obtaining such knowledge is clearly a priority if full benefit is to be made of such areas and management plans made for targeted species.

A major concern today is the likelihood of global change and the effects this will have on protected areas. On the one hand demographic change will put some protected areas at risk in coming decades and on the other hand, some of the predicted effects of climatic change may result in protected areas being wrongly located. The implications that this would have, has yet to be digested by the conservation world.

EX SITU CONSERVATION

The techniques of *ex situ* conservation often form an important and indeed critical element in comprehensive conservation programs although less attention has been paid to them by the conservation community which up to the present has been primarily concerned with habitat protection. Unfortunately habitat loss and fragmentation has reduced many species to populations of small number, often below the size needed for their continued survival without management. In such cases *ex situ* conservation may a necessary component of a recovery strategy or reintroduction plan.

Until recently, *ex situ* conservation has been employed mainly by the genetic resource sector, for crop plants, fodder plants and some tree species, and by botanic gardens which, by definition are concerned with introducing and growing plants away from their natural habitats, although usually not in sufficient quantities to be regarded as a conservation stock. Most agricultural germplasm collections are concerned with conserving a limited range of crop plants and the great bulk of their accessions are of cultivars. Recent studies have drawn attention to the need for more emphasis to be given to the conservation of a germplasm of wild species, including crop relatives, in addition to the current efforts being made by FAO, IBPGR, the CGIAR Crop Centers, IBPGR and other international and national bodies such as the USDA with its National Germplasm System, the Nordic Gene Bank and CENARGEN, Brazil.

Although there have been recent moves by IBPGR to widen
the range of the crops it covers, the total number of species
covered by the international germplasm system on which there
are active programs is about 500 according to Professor J.G.
Hawkes. On the other hand, many of the crop genebanks include
small quantities of seed material of wild species and the
total held is probably in well in excess of 10,000, including
wild relatives of crop species, but few of these are the subject of breeding or selection programs.

Limited germplasm collections of wild species are held by
genebanks such as those of the Royal Botanic Gardens, Kew at
Wakehurst Place, by the Universidad Politecnica de Madrid with
its Crucifer and Artemis collections of Mediterranean species
and by the California Wild Flower Society. Many botanic gardens are in the process of establishing seed banks, having
been involved in conventional means of collecting and storing
material for exchange through the Seed List or Index Seminum
system for many years. In addition the so-called living collections of the world's 1500 botanic gardens and arboreta represent the largest assemblage of plant diversity outside
nature, running into tens of thousands of species.

The diversity of the collections in botanic gardens is
truly amazing and includes thousands of species of orchids of
wild origin, cacti and succulents, bromeliads and many other
important groups (Heywood, 1990). Most of this material
exists as very small accessions, of one or a few individuals
and does not, therefore represent an adequate sample of the
genetic variation in the species concerned to serve as a suitable genetic resource collection. Often a botanic garden
holds material, either in the open or under glass, which is
unique and not held by any other garden. In some cases the
species concerned is even extinct in the wild so that the
botanic garden material is all that remains of that species.
An example is the Royal Botanic Garden, Peradeniya in Sri
Lanka which contains specimens of over 20 species endemic to
the Sinharajah forest region but which have not been recol-
lected during recent detailed floristic surveys of the area.
The problem of what botanic gardens should do when faced with
this kind of conservation situation has not been fully

resolved.

The Botanic Gardens Conservation Secretariat was set up in 1987 by IUCN to encourage and coordinate the conservation activities of botanic gardens and became an independent body in 1990. It consists currently of a network of over 300 member gardens in 67 countries and is the main international organization dealing with the *ex situ* conservation of germplasm of wild plant species. It maintains a database of the conservation holdings of member botanic gardens with currently some 50,000 records of accessions. It published with WWF the Botanic Gardens Conservation Strategy which details the ways in which botanic gardens should undertake the practical conservation of plants and organize their activities. The Strategy estimates that botanic gardens should be able to develop the capacity to maintain conservation collections, mainly in the form of seed, of up to 25,000 species with a view to holding them in medium or long term storage for future use - for breeding, reintroduction programs or as insurance against possible extinction of the species in the wild.

The main forms of *ex situ* conservation of plants are storage of seed under conditions of desiccation and low temperature; field gene banks as in botanic gardens or research stations; clonal collections; tissue or cell culture; and cryopreservation.

SETTING PRIORITIES

Because resources for conservation will always be insufficient to meet all the likely demands for action, it is necessary to establish priorities. These will vary at local, national, regional and global levels. Setting priorities requires information on taxonomic identity, distribution, human uses, status and trends at all levels of biodiversity. Such information is gathered by bodies such as the World Conservation Monitoring Centre, for endangered species of plants and animals, protected areas and wildlife trade, the UNEP Global Environment Monitoring System (GEMS), FAO, IBPGR, The World Resources Institute and by numerous national databases and information systems.

HOTSPOTS

It is estimated that the world's tropical forests contain
about a half of the world's species but even within this biome
a disproportionate amount of the biodiversity is concentrated
in a relatively small number of areas. The U.S. Academy Committee on Research Priorities in Tropical Biology identified
eleven areas deserving of special attention because of their
high diversity, large number of endemics and rate of forest
conversion. These were: coastal forests of Ecuador; the
"cocoa" region of Brazil; Eastern and southern Brazilian Amazon; Cameroon; mountains of Tanzania; Madagascar; Sri Lanka;
Borneo; Sulawesi; New Caledonia and Hawaii.

A different approach was adopted by Myers (1988b) who
identified a series of ten tropical forest 'hot spot' areas
-Madagascar, Atlantic Coast of Brazil, western Ecuador, the
Colombian Choco, the uplands of western Amazonia, eastern
Himalayas, peninsular Malaysia, northern Borneo, the Philippines and New Caledonia. These occupied an area of 292,000
km^2 (or 0.2% of the earth's land surface) and comprised only
3.5% of primary tropical forests, yet contained an estimated
34,400 endemic species, amounting to 27% of the total of all
tropical forest plant species or 13.8% of the earth's plant
species. He later extended this type of analysis to focus on
another eight areas, four of them in the tropics forests and
four in Mediterranean-type zones. The former contain at least
2,835 endemic species in 18,700 km^2 or 1.1% of the earth's
plant species in 0.013% of the earth's surface, while the four
Mediterranean type areas contain 12,720 species or 5.1% of the
earth's plant species in 0.3% of the earth's surface. The aim
of identifying these areas was to allow conservation efforts
to be concentrated where the pay-off would be greatest.

CENTERS OF PLANT DIVERSITY

A more critical and complex approach to the identification
of priority areas for plant conservation is the IUCN Centers
of Plant Diversity project supported by WWF, ODA and EEC. Its
aim is to identify those areas of high plant diversity considered to be of global significance and will be a guide to the
most cost-effective and practical ways to conserve as many

plant species as possible in a selected number of areas.
Emphasis will be given to the potential value of each site
selected for the conservation and sustainable development of
its plant resources and the likelihood of its ability to survive in the face of demographic pressure.

The criteria for selection are: first, that the area is
evidently species-rich, even though the number of species may
not be accurately known; secondly, the area is known to contain a large number of plant species endemic to it. In addition the following characteristics will be taken into account:
a) the site contains important gene pools of plants valuable
to humans or of potential value; b) the site contains a diverse range of habitat types; c) the site contains a significant proportion of species adapted to special edaphic conditions; and d) the site is threatened or under imminent threat
of development.

It is expected that approximately 200 sites will be chosen. This is a much more difficult and demanding task than
choosing a small number which are virtually self-selecting.
As part of the process, Regional Overviews will be prepared
summarizing the floristics, endemism, vegetation and plant
resources of the major regions of the world. These Overviews
will list all the major candidate botanical sites of the world
although full Data Sheets will be prepared for only a selection of these, i.e. the 200 Centers of Plant Diversity.

The project is being coordinated at the IUCN Plants Office
and involves extensive cooperation with many organizations and
individuals around the world. National committees have been
set up in several cases to help with the selection of sites
and preparation the Data Sheets.

Centers of Plant Diversity, which is planned to be published as a 3 volume compendium in 1992, will provide governments, aid agencies and conservation organizations with a
global perspective on *in situ* priorities for plant conservation and will provide the basis for national and regional
strategies and action plans for conserving plant diversity.
It will help set priorities for the selection of additional
protected areas and for botanical inventories. Each Data
Sheet will include an outline Conservation Strategy and it is

expected that these will be developed locally as the Project
moves on to the implementation phase. CPD is a means to an
end and as in all conservation projects, the ultimate aim is
to achieve conservation on the ground.

RAPID ASSESSMENT PROGRAM (RAP)

     A recent new approach to help decide on tropical conserva-
tion priorities is the Rapid Assessment Project (RAP) whose
aim is to gather data quickly that will help decide which
tracts of endangered rain forest are most deserving of protection. The work is undertaken under the auspices of Conservation International and a small team has carried out rapid
assessments in northern Bolivia and western Ecuador. The team
goes to remote, unsurveyed sites and using satellite imagery,
aerial reconnaissance and field surveys, produces inventories
in a matter of weeks rather than months or years as is custo-
mary. It is a very limited, "quick and dirty" approach that
ignores other aspects such as economics and social values and
current uses of the forests but it has already come up with
some interesting results. For example, in the Andean footh-
ills of northern Bolivia, it surveyed Alto Madidi which it
found to be one of the most diverse rain forests in the world.
The botanist Al Gentry found 204 woody species within a quarter of an acre, many of them new to Bolivia and some new to
science.

CONCLUSIONS

     Enormous amounts of money and energy are being invested in
attempts to conserve the rich diversity of plant and animal
life in the world's tropical ecosystems. Yet the rate of
habitat loss and consequent endangerment of plant and animal
populations continues unrelenting. The Tropical Forestry
Action Plan, despite wide support, has not led to any diminu-
tion in the loss of tropical forests and indeed the rate of
conversion has increased in recent years since the plan was
initiated. There is a need, therefore, to act quickly but in
a coordinated manner, so that the resources available can be
deployed as effectively as possible. Three factors that will
help us work more effectively are information, coordination

and involvement of local communities. We need to know what there is to conserve and where - inventories, centers of diversity; joint action using integrated approaches which involve *in situ* and *ex situ* techniques as appropriate; and finally we must not overlook the often very detailed knowledge and understanding of ecosystems and species and their uses possessed by local inhabitants of these areas in the tropics. We must learn to work with such groups and not impose inappropriate technologies on them that not only replace local biodiversity but change their social and economic systems and ways of life. The title of the new IUCN document that will succeed the World Conservation Strategy is "Caring for the World" and the only effective way of doing so is by working together.

Literature Cited

Falk, D. A. 1990. Integrated strategies for conserving plant genetic diversity. *Ann. Missouri Bot. Gard.* 77: 38-47.

Heywood, V. H. 1990. Botanic gardens and the conservation plant resources. *Impact of Science on Society* No. 158: Pp. 121-132.

McNeely, J. A., K. R. Miller, W. Reid, R. A. Mittermeier and T. Werner. 1989. *Conserving the World's Biological Diversity.* World Resources Institute/International Union for Conservation of Nature and Natural Resources/Conservation International/World Bank, Washington D. C. and Gland, Switzerland.

Myers, N. 1988a. Threatened biotas: "hot-spots" in tropical forests. *The Environmentalist* 8(3): 1-20.

_____. 1988b. Tropical forests and their species. Going, going...? Pp. 27-35. In: E. O. Wilson and F. M. Peter (eds.), *Biodiversity.* National Academy Press, Washington D. C.

Raven, P. H. 1987. The scope of the plant conservation problem world-wide. Pp. 10-19. In: D. Bramwell, O. Hamann, V. Heywood and H. Synge (eds), *Botanic Gardens and the World Conservation Strategy.* Academic Press, London.

_____. 1988. Our diminishing tropical forests. Pp. 119-122. In: E. O. Wilson and F. M. Peter (eds.), *Biodiversity.*

National Academy Press, Washington D. C.

Simberloff, D. 1986. Are we on the verge of a mass extinction in tropical rain forests? Pp. 186-190. In: D. K. Elliott (ed.), *Dynamics of Extinction*. John Wiley and Sons, New York.

Wilson, E. O. 1988. The current state of biological diversity. Pp. 3-18. In: E. O. Wilson and F. M. Peter (eds.), *Biodiversity*. National Academy Press, Washington D. C.

THE GENE LIBRARY - PRESERVATION AND ANALYSIS OF GENETIC DIVERSITY IN AUSTRALASIA

John S. Mattick
Effie M. Ablett
and
Daryl L. Edmonson

Centre for Molecular Biology and Biotechnology
The University of Queensland, Brisbane 4072, Australia.

Summary - The University of Queensland established a Gene Library in 1989 to collect and preserve DNA from Australasian species which are rare, endangered or have bio-technological value. Preserving DNA is about preserving information. The genetic content of different species varies between about 1 million and 10 billion base-pairs, and with an estimated 10 million species on earth, the genetic database holds about 10^{15} bits of information. The introduction of gene libraries to house source material and to provide databases is therefore a logical step. We anticipate that such libraries will be established in key locations around the world over the next decade, to serve two functions: (a) a resource for exploring biological diversity and evolutionary history; and (b) a resource of increasing importance for the development of biotechnology. There is considerable urgency in establishing such collections, with the accelerating destruction of habitats and consequent loss of species. Despite increased awareness of the need for habitat preservation, extinction of some species appears inevitable. It is imperative that we collect genetic information from as wide a variety of existing species as possible.

The Gene Library in Brisbane currently holds approximately 500 specimens, and many more thousand specimens have been pledged from museums, herbaria, and wildlife research groups. The National Parks and Wildlife Service assists by supplying monthly lists of researchers with permits to collect fauna. The collected material (normally blood, liver, or leaves) is stored in three forms: (a) whole cells/tissue (either frozen

or desiccated); (b) isolated DNA; and (c) cloned genomic
libraries. The latter are available to other research groups,
for example for preparing DNA probes for use in conservation
research. Techniques have been developed to store blood or
tissue in the field (with or without freezing facilities) for
the best recovery of DNA, and to extract DNA from samples
which are often small and occasionally lysed. The Library
uses an Applied Biosystems automated DNA extractor which gives
good quality DNA by virtue of the large volumes of solvents it
uses, and this service is available to other research groups.

One obvious benefit of the collection, apart from its
scientific value, is the generation of a resource for gene
technology. Very little of the Earth's genetic diversity has
been properly explored, let alone accessible from a central
collection. This diversity includes unique developmental features, adaptations to specialized environments, and unusual
biochemistries involving complex organic synthesis and degradation of bioactive compounds. In the area of plants alone,
only about 2-3% of angiosperms have been examined for their
pharmaceutical potential, and a much smaller fraction for
their broader genetic potential as donors of useful genes for
industry or agriculture. The Gene Library is intended to be
both a heritage collection and a resource for science and
technology. Ultimately it may also have the benefit of speed-
ing the integration of molecular genetics with whole organism
biology, perhaps the true end point of biological research.

INTRODUCTION

In Australia, there are many plants, animals, insects, and
microorganisms which differ significantly from those of the
rest of the world. These, and wild species generally, are
sources of unique genes which may give new insights to biological mechanisms and evolution, as well as having potential
uses in agriculture, medicine and conservation biology. The
Gene Library has been established firstly to preserve the
genetic information of Australasian species by isolating and
storing their DNA; and secondly, to turn this into an active
resource for science and technology, by making it readily
available to researchers throughout the world.

AUSTRALIA'S UNIQUE FLORA AND FAUNA

Since the break-up of Gondwanaland, Australia's isolation from other landmasses has resulted in a great diversification and radiation to produce many species, and even whole families of plants and animals unlike anything else in the world. In particular, sclerophyll plants such as the eucalyptus, banksias, and hakeas evolved in response to the increasingly arid climate (Johnson and Briggs, 1975). Many new species of marsupials developed with alterations in environmental conditions.

Today, Australia contains some of the world's richest reservoirs of biological diversity, including the Great Barrier Reef, tropical and sub-tropical rainforests, and the semi-arid and arid grasslands. The rainforests of North Queensland rank with those of Amazonia and Madagascar for their biological richness and species diversity (Keto, 1986).

As well as the evolution of new biota, Australia is notable for having some very ancient species and relic families which failed to survive elsewhere in the world. These include the monotremes (egg laying mammals) represented by the platypus and the echidna. The climatic cooling that occurred in the latter Tertiary and early Oligocene was responsible for the elimination of many organisms that were warm adapted. At this time the Australian plate's northern migration into warmer regions counteracted the extreme cooling that occurred in other continents (Kemp, 1981). As a result, a wide array of primitive Gondwanaland biota survived, particularly in tropical rainforest environments. During the Pleistocene glacial periods (up to 10,000 years ago), the climate became cooler and drier. The tropical rainforests contracted into small moist areas. Today these refugia are the home of several species of plants and insects which have survived more or less unchanged for over 100 million years.

The world's highest concentration of primitive angiosperm families are found in the refugia of the North Queensland rainforests (Keto, 1986). These remnants are all that is left of the southern evolutionary stream of angiosperms, and differ markedly from other ancient angiosperms found elsewhere in south-east Asia (Schuster, 1972). Notable examples are the

endemic monotypic families - Idiospermaceae and Austrobaileyaceae, and the Australasian endemic ditypic families - Eupomatiaceae and Himantandraceae. Morphological studies have also
numbered these unique Australian species among the most
ancient of the living angiosperms (Endress and Honegger, 1980;
Endress, 1984).

THE LOSS OF AUSTRALIAN SPECIES

Since the first European settlers came to settle this
strange harsh land 200 years ago, they have done everything
possible to obliterate the Australian bush and replace it with
a more familiar European agricultural landscape. Three-
quarters of the rainforests have been cleared and burned.
Two-thirds of the original tree cover has been removed, and
with the consequence that more than half of the arid lands are
now degraded by erosion and salinity. Native plants now com-
pete with foreign species, and introduced animals prey upon
the native fauna. Since 1788, at least 100 plant and 28 animal and bird species have become extinct. Hundreds more hover
on the brink of extinction, including the Northern Hairy Nosed
Wombat, Bennett's Tree Kangaroo, the Greater Bilby, the
Bridled Nail-tailed Wallaby, the Glossy Black Cockatoo, the
Ground Parrot, the Western Swamp Turtle (the world's most
endangered reptile), and many *Eucalyptus* species.

Australia's high rate of habitat destruction is one of the
major issues on the national agenda for the 1990's, and every
effort is being made to preserve the unique species and their
environment. Sadly, however, many of the endangered species
have suffered such a reduction in numbers that their gene
pool is now too narrow for long term survival: extinction
appears inevitable. Queensland herbarium studies indicate
that at least 20 Queensland plant species are likely to become
extinct within ten to twenty years (Thomas and McDonald,
1989). It is essential that we preserve something of these
species before they are lost forever.

THE GENE LIBRARY

To address Australia's loss of biological diversity, the
Centre for Molecular Biology and Biotechnology at the Univer-

sity of Queensland established the Gene Library in 1989. In
setting up this collection of genetic material from Australia's unique flora and fauna, first priority has been given
to samples from species which are rare, endangered or have
biotechnological value. The latter category will include
samples of commercial crops and livestock and unusual variants.

Samples of tissues and cells are stored either cryogeni-
cally or desiccated depending on the type of sample and organism involved. From these, enough DNA is extracted for cryog-
enic storage of 'pristine' samples at two locations and to
prepare genomic libraries, which will be available for dissemination to other research groups. The Gene Library also
plans to assist researchers in the preparation of gene probes
DNA sequencing. In the long term, it is envisaged that the
collection will become a comprehensive data base of DNA
sequences, as this information becomes available.

SOURCES OF TISSUE SAMPLES

Close links have been developed with existing collections
of biological material, particularly museums, herbaria, and
wildlife research groups. The establishment of the Gene
Library has received considerable support from these groups,
who hold hundreds of thousands of samples as potential sources
of DNA. A list of collaborating institutions is given in
Table 1.

The Queensland National Parks and Wildlife Service assists
by supplying monthly lists of researchers with permits to collect fauna. Samples have largely been obtained from existing
collections, and from researchers who are collecting samples
as part of their own studies. In this way, there is minimal
drain on wild gene pools, which can be crucial for endangered
animal species, and in addition, samples are identified by
experts in their field. Samples are of little use if they are

Table 1. Collaborating Institutions: The following organizations have given us access to their collections, or are collecting material for us during their field studies.

Australian Collection of Bacteria
Australian Museum
Australian National Wildlife Collection
Currumbin Sanctuary
Departments of Entomology, Zoology, Botany, Parasitology, and
    Farm Animal Medicine and Production at the University of
    Queensland
Department of Genetics, La Trobe University, Melbourne
Queensland Herbarium
Queensland Museum
Queensland National Parks and Wildlife Service
South Australian Museum
Taronga Park Zoo
Western Australian Wildlife Research Centre

not correctly identified. A detailed specimen information
sheet (Table 2) is filled out for each sample, containing
information on how the specimen was identified, by whom and
the criteria used for identification. When whole or part of
an animal specimen is preserved elsewhere, this is documented
to allow one to return to the original specimen at a future
date to confirm identification, or to obtain additional material from the same individual. Most plant samples are taken
directly from dried herbarium samples, allowing future access
to the original specimens.

   The collection currently holds about 500 specimens, and is
growing daily. A list of key species held in the Library is
given in Table 3 (see also Figure 1).

Table 2. Information criteria for Gene Library samples: The
following criteria are the basis of a sample information form
filled out for each incoming sample.

1.1 ACCESSION NUMBER
1.2 DONOR INSTITUTE
    1.2.1 Donor identification No.
1.3 SCIENTIFIC NAME
    1.2.1 Genus; 1.2.2 Species; 1.2.3 Subspecies;
    1.2.3 Authority
1.4 SPECIES IDENTIFICATION
    1.2.1 Identified by; 1.2.2 Identifying characteristics
1.5 COMMON NAME
1.6 ENDANGERED STATUS
2.1 COLLECTOR'S NAME
2.2 COLLECTING INSTITUTE'S SPECIMEN No.
2.3 COLLECTION DATE
2.4 COLLECTION LOCATION
    2.4.1 Country; 2.4.2 State; 2.4.3 Location of Collection
    site (grid reference if known)
    2.4.4 Altitude of Collection site
2.5 HABITAT TYPE
3.1 SAMPLE DETAILS
    3.1.1 Tissue type; 3.1.2 Quantity; 3.1.3 Number of tubes
    3.1.4 Method of preservation; 3.1.5 Condition of donor
    organism
3.2 ANIMAL SPECIES
    3.2.1 Age; 3.2.2 Sex; 3.2.3 Pedigree; 3.2.4 Other descrip-
    tors;
    3.2.5 Is part or all preserved elsewhere? If so, where?
3.3 PLANT SPECIES
    3.3.1 Growth type; 3.3.2 Height; 3.3.3 Description of
    flowers; 3.3.4 Description of fruit; 3.3.5 Description
    of bark; 3.3.6 Other descriptors

Figure 1. Key species held by the Gene Library: (a) *Austro-baileya scandens*. This species is the only living represent-ative of the Gondwandic family, Austrobaileyceae, which may

occupy an isolated systematic position between two orders of
plants (Endress and Honegger, 1980). It is endemic to North-
East Australia. (b) "Drip Tips". A new species (as yet
unidentified) of a monotypic genus. Only one population is
known in the World Heritage area of Australia's wet tropics.
(c) *Davidsonia* sp. nov. An endangered Gondwandic species
restricted to central Eastern Queensland. Davidsoniaceae is a
monogeneric family with 2 species, allied to the Cunoniaceae
of the southern hemisphere. (d) *Fontainea vendsa*. An endangered species restricted to South-East Queensland. The Gondwandic genus is restricted to Australasia and New Caledonia.

Table 3. Key species held by the Gene Library: Samples of the
following species are considered important as a consequence of
their endangered status and biological significance (details
given).

Mammals	Endangered Status and Biological Significance
Humpback Whale#	Drastically reduced
Pygmy Sperm Whale#	Status obscure, surveys required
Spotted-Tailed Quoll	Threatened
Bridled Nailtail Wallaby	Very rare, decreasing in numbers
Ghost Bat	Severely depleted, numbers continuing to decline
Brush Tailed Bettong	Threatened, very restricted distribution
Snow Leopard#	Endangered, possibly only several hundred individuals remaining
Echidna	Australasian endemic sub-class (the monotremes, egg laying mammals)
Greater Bilby	One of Australia's most highly endangered species
Northern Hairy Nosed Wombat*	Only 65 individuals remain

Birds

Ground Parrot	Rare
Goldern Shouldered Parrot	Endangered

Table 3 (continued)

Bush Stone Curlew	Severely depleted, numbers continuing to decline
Night Parrot*	Thought to be extinct until recent find
Squatter pigeon	Severely depleted and continuing to decline. Susceptible to introduced predators
Spotted Bower Bird	Vulnerable. Range greatly reduced

Reptiles

Diamond Python	Numbers declining, subject to substantial export trade
Loggerhead Turtle#	Severely depleted
Fresh Water Crocodile	Vulnerable
Green Turtle#	Severely depleted
Green Python	Vulnerable, subject to substantial export trade
Western Taipan	Threatened
Gastric Brooding Frog*	First found in 1973, now presumed extinct due to habitat destruction. Its reproductive habits are unique: tadpoles develop in the female's stomach.
Western Swamp Turtle*	The world's most endangered reptile: less than 50 individuals remain

Plants

Acanthaceae

Asystasia Between *gangetica* and *australasica*	Very rare, species known only from the type collection
Brunoniella spiciflora	Endangered, occurring in small populations
Graptophyllum excelsum	Endangered, distribution less than 100 km

Table 3 (continued)

Hemigraphis royenii	Rare, distribution less than 100 km.
Isoglossa eranthemoides	Vulnerable, occurring in small populations
Lepidagathis sp.	Rare, distribution less than 100 km.
Xerothamnella herbaceae	Endangered, restricted to less than 100 km

Agavaceae

Cordyline congesta	Rare, restricted to within 100 km
Cordyline fruticosa	Poorly known species, restricted to 100 km

Amaranthaceae

Ptilotus remotiflorus	Poorly known species, restricted to 100 km

Annonaceae

Ancana stenopetala	Primitive Angiosperm, Australian endemic ditypic genus
Ancana hirsuta	Primitive angiosperm, Australian endemic ditypic genus
Haplosticanthus spp.	Primitive angiosperms, Australian endemic genus, very restricted
Polyaulax sp.	Primitive angiosperm, poorly known species, restricted to 100 km
Melodorum sp. Q1	Primitive Angiosperm, poorly known species, restricted to 100 km

Apocynaceae

Parsonsia densivestita	Poorly known species, restricted to 100 km

Araliaceae

Polyscais bellendenkeriensis	Vulnerable, restricted to 100 km

Asclepiadaceae

Marsdenia coronata	Endangered, restricted distribution
Marsdenia longiloba	Endangered, restricted distribution

Table 3 (continued)

Austrobaileyaceae
Austrobaileya scandens   Gondwandic, monotypic family

Davidsoniaceae
Davidsonia sp. nov.   Gondwandic, monogeneric family with
                                  2 species, allied to the *Cuno-
                                  niaceae*, endangered

Euphorbiaceae
Euphorbia carissoides   Presumed extinct
Fontainea vendsa   Gondwandic genus, threatened,
                                  restricted distribution

Eupomatiaceae
Eupomatia bennettii   Primitive angiosperm, Australasian
                                  endemic ditypic family
Eupomatia laurina   Primitive angiosperm, Australasian
                                  endemic ditypic family

Hernandiaceae
Hernandia bivalvis   Vulnerable, occurring in small popu-
                                  lations, primitive angiosperm
Valvanthera albiflora   Potentially vulnerable, occurring in
                                  small populations, primitive
                                  angiosperm

Hydrocharitaceae
Hydrocharis dubia   Vulnerable, distribution less than
                                  100 km

Idiospermaceae
Idiospermum australiense   Gondwandic, monotypic family, vul-
                                  nerable, occurring in small
                                  populations

Table 3 (continued)

Lamiaceae

Plectranthus argentatus   Potentially vulnerable, occurs in
                                     small populations

Teucrium ajugaceum   Little information on distribution,
                                     possibly a range of less than
                                     100 km

Westringia rupicola   Potentially vulnerable, distribution
                                     less than 100 km

Westringia parvifolia   Vulnerable, distribution less than
                                     100 km

Plectranthus arenicola   Endangered, occurs only in a 10 m x
                                     10 m area at site of proposed
                                     space port

Plectranthus argentatus   Potentially vulnerable, occurs in
                                     small populations

Lauraceae?

"Drip Tips"   New species as yet unidentif-
                                     ied, monotypic genus, endan-
                                     gered, occurs in less than 0.5
                                     Ha

Leguminosae

Acacia porcata   Endangered, only 6 plants remain in
                                     an area of 0.1 Ha

Monimaceae

Dryadodaphne Q1   Rare, restricted to 100 km
Wilkiea wardellii   Rare, restricted distribution

Myrtaceae

Decaspermum sp.   Endangered, known only from one
                                     tree

Table 3 (continued)

Musaceae

*Musa jackeya**   Wild banana
*Musa fitzlanif**   Wild banana

Poaceae

Garnotia stricta   Potentially vulnerable, occurs in
                                    small populations
*Triunia robusta**   Endangered, 50 plants remain in two
                                    populations

*Specimens pledged or in transit
Not endemic to Australasia

SAMPLE PROCESSING

The Library uses an Applied Biosystems model 340A DNA
extractor which gives good quality DNA by virtue of the large
volumes of solvents it uses. It is also relatively reproducible, as automation results in reduced variation of the extraction technique. The Applied Biosystems isolation procedure
relies on Proteinase K digestion in 4M urea with sodium laur-
ylsarcosine, followed by phenol/chloroform and chloroform
extractions. The DNA is precipitated by addition of ammonium
acetate and isopropanol and collected onto a filter. The
standard machine protocols work well for crude nuclear preparations from blood, liver and fresh plant specimens. We have
found the 'whole blood' protocols useful in obtaining high
yields from samples which are often small and occasionally
lysed. We are presently adapting these techniques to the isolation of DNA from dried herbarium samples. Doyle and Dickson
(1987), Pyle and Adams (1989) and Adams *et al*. (this volume)
advocate desiccation as the best method for preserving plant
samples that are stored prior to DNA extraction.

Blood samples from non-mammalian species with nucleated
red blood cells are a convenient source for DNA isolation. As
little as 0.2ml will provide ample DNA for our needs (400µg).
This can be frozen as soon as possible after collection (with

the addition of EDTA as an anticoagulant), or, if the samples
have to be stored at room temperature, these can be added to
the Applied Biosystems Lysis Buffer (Galbraith, 1989). We

found no reduction in yield of restrictable, high molecular
weight DNA when samples were stored for 4 days in lysis buffer, even at 37°C.

As mammalian species require relatively large volumes of
blood to obtain useable quantities of DNA, liver samples are
used whenever available. Routinely, 0.5g is frozen and can
yield up to 800µg DNA. For storage at room temperature, the
liver is placed in APS medium (Arctander, 1988). This contains NaF and EDTA to inhibit nucleases, and thymol to prevent
bacterial and fungal growth. APS may be also useful for storing fresh plant specimens but has not yet been tried in this
context. Although there is a significant reduction in DNA
yields from liver samples stored at room temperature, or at
37°C (70% and 82% respectively, after 7 days), the DNA is of
high molecular weight, cuts well with restriction enzymes, and
sufficient can be obtained for our needs from 1g of liver.
Freshly frozen mammalian blood is processed similarly to
nucleated red blood cells. Storage of room temperature
samples in Lysis Buffer is impractical, as a sufficient dilu-
tion to prevent coagulation is too low to precipitate the DNA.
Other methods for ambient temperature storage (which may be as
high as 35°C in the field) of mammalian blood are presently
being investigated. Many Australian marsupials and bats are
too small to obtain the volumes of blood required for at least
200µg DNA (6ml), even with repeated bleeding. In these cases,
we store smaller amounts of DNA from several individuals.

Purified DNA is redissolved in TE buffer (10mM Tris-Hcl,
0.1mM EDTA pH 8.0), with the addition of 5µl chloroform per
ml. After a sample is taken to determine concentration and
purity by U.V. absorption spectra, it is aliquoted and stored
at -80°C. In most cases, we have not yet had the opportunity
to prepare cloned genomic libraries, but we envisage that they
will be prepared using partial *Sau*3A or *Pst*I digestion to
generate quasi-random overlapping clones. At this stage standard lambda phage vectors will be used. In the future, we may

attempt to make libraries with larger DNA fragments, using YAC
(yeast artificial chromosome) or Bacteriophage PI as cloning
vectors.

CONTAMINATION BY FOREIGN DNA

As the DNA is to be amplified in plasmid libraries, it is
important that samples have minimal contamination with microbial and human DNA which may be present *in situ* when samples
are collected, and/or introduced during processing. Glassware
and equipment used in the Gene Library are rinsed in 1N HNO_3
whenever necessary, to depurinate any contaminating DNA. Dis-
secting instruments, and forceps used for handling purified
DNA filters are treated to prevent cross-contamination between
samples. Gloves are used to prevent contamination from skin
cells, and herbarium samples are soaked in Lysis Buffer to
remove surface microbial flora. As an additional precaution,
purified DNA or cloned gene libraries may be subsequently
screened with suitable probes, to give users an estimate of
contamination.

FUTURE CONSIDERATIONS

With the dissemination of genomic DNA libraries, the DNA
of the Gene Library could become analogous to the cells of the
American Type Culture Collection. This collection has stimulated biomedical research in two ways:

Firstly, when raw materials are readily available, considerable new research is generated. The availability of cloned
genes will facilitate DNA sequencing studies and the development of DNA probes. These tools are revolutionizing basic
whole organism studies, as well as applied research. Species
specific DNA probes are being used increasingly for species
identification. DNA probes directed against polymorphic sites
in the genome provide new means for analyzing the genetic diversity of populations in the wild. Such information is crucial to the development of rational conservation strategies.
Hopefully, the use of DNA from the Gene Library will generate
many more such applications of DNA technology in wildlife
research and management in the near future. DNA sequencing is
now being used to explore biological diversity, determine phy-

logenetic relationships and define speciation, as well as giving some insights into evolution.

Secondly, the use of the same cell lines by researchers
throughout the world has allowed the work of many research
teams to come together to form a complete picture of standard
models. In the same way the use of a genomic DNA library from
one individual will standardize work on that organism, especially in comparison of DNA sequences.

THE INTERNATIONAL PERSPECTIVE

As the use of molecular biology in whole organism studies
increases, we anticipate that collections such as ours will be
established in key biogeographical locations around the world.
These 'nodes in a DNA Bank-Net' will serve two functions:
firstly as a collection of genetic heritage and a resource for
the exploration of biological diversity and evolutionary history, and secondly as a resource of increasing importance for
the development of biotechnology.

It is becoming increasingly obvious that molecular genetics is ultimately an information science. DNA is biological
software. The DNA content of different cells varies over
about five orders of magnitude from primitive prokaryotes to
the higher plants. One can attempt some rough calculations of
the total size of the genetic database which exists in the
biosphere. Assuming that there is presently somewhere around
10^7 different species (depending on how much allowance is made
for diversity in complex ecosystems), with an average coding
content of about 10^8 base pairs of DNA, then the total database is around 10^{15}, give or take an order of magnitude or so.
This is a very large number, although perhaps something of an
overestimate, since no discounting has been attempted for
homologous genes among related species. Nevertheless, when
one considers that the total sequence data presently available
is around 5×10^7, then we have hardly yet scratched the surface.

There are two further points to be made about this database. It is arguably the most important on the planet. It is
also presently under serious threat from human agricultural
and industrial expansion, resulting in not only local habitat

destruction but possibly more widespread and regional damage from climatic change or human conflict. It is vital that we take urgent steps to preserve this database, not only by environmental protection, but also by the collection and storage of cells and DNA samples from endangered species. It appears certain that species will continue to be lost in the near future, at least until some balance is restored to man's interaction with the natural world. The genetic information in these organisms is our biological heritage, the product of millions of years of evolution, and is as priceless as great works of art or literature. Preservation of such material would at least allow the retention of this information for future analysis, and possibly even reactivation as cell and molecular technologies improve.

The second point to be made is that it is the extent to which the genetic database has been characterized will be rate-limiting in the development of biotechnology. This is rarely acknowledged but, with few exceptions, the successful applications of genetic engineering technology have been predicated on a very good knowledge of the biological system involved, and it seems clear that biotechnological possibilities will accelerate exponentially as information is linearly extracted from the database. This is particularly true for the higher plants and animals.

The technology for obtaining DNA sequence data is developing rapidly, and there are now automated DNA sequencers in widespread use. This development is being accelerated by the Human Genome Project, and associated studies of key organisms in the evolutionary hierarchy. One would expect most biological research laboratories to embrace DNA sequencing relatively quickly, if they have not done so already, as an important part of their normal activities. The recent introduction of polymerase chain reaction technology, which is revolutionizing methods for genetic manipulation and analysis, including diagnostic tests, will create an increased demand for DNA sequence information, and promote the lateral extension of the database to other organisms.

THE NEED FOR IMMEDIATE ACTION

There is considerable urgency in the establishment of collections of source material to preserve genetic information, in the face of the accelerating destruction of habitats and species loss. Losing a species without collecting a genetic specimen is like burning the last copy of Hamlet without taking a photocopy.

Despite the increased awareness of the need for habitat preservation, the main issue is not so much whether there will be significant loss of species, but rather the extent to which this can be prevented. Given that the Earth's population will at least double, and probably triple, in the next 75 years (even if widespread measures to limit growth were enacted immediately), the local and global pressure on our ecosystems will be severe, if not intolerable. While habitat preservation may be the immediate goal of those concerned with the preservation of biological diversity, population growth with its agricultural and industrial consequences remains the underlying problem, and will not be solved in the short term. Although strenuous efforts to preserve habitat may limit the damage, there seems little doubt that some extinction is inevitable, even without considering the potential for more apocalyptic problems arising from possible conflict or climatic change. For all of these reasons, it would seem prudent to begin to collect genetic information from as wide a variety of existing species as possible, as a heritage collection and insurance against loss.

THE SCIENTIFIC VALUE

Even in the absence of immediate threats to wild genetic stocks, it is clear that genetic material should be gathered and analyzed as a scientific venture in its own right. Genetic information is becoming increasingly accessible with the rapid development of DNA cloning and sequencing technologies. While much attention is being paid to key model organisms such as *E. coli*, yeast, *Arabidopsis*, *Caenorhabditis*, *Drosophila*, mouse and man, we should not ignore the wealth of information in other species. The scientific value of new genes from wild species has yet to be realized. The establishment of gene

libraries and DNA Bank-Net will do much to assist this process, and provide a link between the raw material and the
deciphering of its information. This should not be limited to
plants, but extended to cover all kingdoms.

BIOTECHNOLOGY

One obvious benefit of such a collection, apart from its
scientific value, is the generation of a major resource for
gene technology. Very little of the genetic diversity in the
biosphere has been properly explored, let alone accessible
from a central collection. This diversity includes not only
unique developmental features, but also a range of physiological adaptations to specialized environments (for example in
thermophilic, psychrophilic, halophytic, or xerophytic organisms) and unusual biochemistries involving complex organic
synthesis and degradation of bioactive compounds. In the area
of plants alone, it has been estimated that only 2-3% of
angiosperms have been explored for their pharmaceutical potential, and a much smaller fraction for their broader genetic
potential as donors of useful genes for industry or agriculture. The primitive angiosperms may prove particularly useful
in this context. The emergence of the first angiosperms
necessitated fundamental changes in plant physiology. The
primitive angiosperms may have rudimentary molecular systems
which could fill evolutionary gaps and help our understanding
of their more complex modern counterparts. For example, genes
determining the biochemistry, physiology, and development of
their reproductive systems may be invaluable to modern plant
biotechnology.

A reservoir of genetic material from a wide variety of
sources will become a major resource for biotechnological
development as the intellectual and technological framework of
molecular genetics develops and more sophisticated algorithms
for predicting structure - function relationships in proteins
come into place. Site-directed mutagenesis adds new and virtually infinite dimensions to genetic manipulation, but we
have yet to properly explore natural diversity.

Literature Cited

Arctander, P. 1988. Comparative studies of avian DNA by
    restriction fragment length polymorphism analysis: Conve-
    nient procedures based on blood samples from live birds.
    J. Ornith. 129: S. 205-216.

Doyle, J. J. and E. E. Dickson. 1987. Preservation of plant
    samples for DNA restriction endonuclease analysis. *Taxon*
    36: 715-723.

Endress, P. K. 1984. The role of inner staminodes in the
    floral display of some relic Magnoliales. *Plant Syst.*
    Evol. 146: 269-282.

_____ and R. Honegger. 1980. The pollen of the Austrobailey-
    aceae and its phylogenetic significance. *Grana* 19:
    177-182.

Galbraith, D. A. 1989. Blood storage in a commercially avail-
    able lysis buffer. *Fingerprint News* 1: 6.

Johnson, L. A. S. and B. G. Briggs. 1975. On the Proteacea-The
    Evolution and Classificaton of a Southern Family. *Bot. J.*
    of the Linnean Soc. 70: 83-182.

Kemp, E. M. (A. Keast, ed). 1981. *Tertiary palaeogeography and*
    the evolution of Australian climate in ecological biogeog-
    raphy of Australia. Pp. 31-50. Junk, The Hague.

Keto, A. 1986. *Tropical rainforests of North Queensland-Their*
    conservation significance. Australian Heritage Commission
    Special Australian Heritage Publication Series, No. 3.

Pyle, M. M. and R. P. Adams. 1989. *In situ* preservation of DNA
    in plant specimens. *Taxon* 38: 576-581.

Schuster, R. M. 1972. Continental movements, 'Wallace's Line'
    and Indomalayan - Australasian dispersal of land plants:
    Some eclectic concepts. *Botanical Review* 38: 3-86.

Thomas, M. B. and W. J. F. McDonald. 1989. *Rare and threatened*
    plants of Queensland. 2nd Edition. Queensland Department
    of Primary Industries Information Series QI88011.

FLORISTIC DIVERSITY, BOTANICAL EXPLORATION AND THE ESTABLISHMENT OF A GERMPLASM BANK FOR CONSERVATION IN THE NEOTROPIC: THE COLOMBIAN VIEW

L. Atehortua

Departamento de Biologia

Universidad de Antioquia

A.A. 1226 Medellin-Colombia

Summary - Colombia, considered as a biotic reserve with a great megadiversity, is also facing the danger of being a country of a great megaextinction, unless some conservation strategies, together with serious political and economical measures, are adopted. *In situ* conservation seems to be far away from the socio-economic reality; therefore, it is proposed *ex situ* conservation be established through germplasm banks as an additional reliable alternative before the extinction of many plant species.

INTRODUCTION

Colombia, with a continental area of 1,141,768 sq. km., is the fourth largest South American country and the second one in biological resources after Brazil.

Quoting Peter Raven (1987), "more than 40,000 plant species-a quarter of the world's diversity occurs in the three Andean countries of Colombia, Ecuador and Peru combined, countries in which the plants, animals and micro-organisms are the most poorly known assemblage on earth". Such biodiversity is the source of germplasm. However, without a systematic inventory, that germplasm loses its significance. A systematic inventory provides the means of scientific communication and is the main framework for most comparative biological studies (Miller, *et al.*, 1989). Prance (1977) estimates 45,000 to 55,000 plant species for Colombian flora, with nearly 13,000 endemic plant species, without taking into account other types of organisms. Among the taxonomic groups, orchids, bromeliads, palms and ferns are the most representative part of our flora. Most of these species are confined to the tropical rainforest, well represented by Choco region and upper Amazon.

Areas rich in endemic species are Sierra Nevada de Santa
Marta, the Guajira Peninsula, La Macarena and many parts of
the Andes. Above all, the Choco region in the western part of
Colombia is the wettest and possibly the richest rain forest
in the Neotropic, with both high diversity and high endemism
(Gentry, 1982).

Although floristic inventory work of Colombia is still far
from being complete, some regional inventories have already
shown its great diversity. According to the plant checklist
for Choco (Forero and Gentry, 1989), nearly 4000 plant species
have been reported. Since the middle of 1986, with the financial support of COLCIENCIAS and the collaborative participation of the Missouri Botanical Garden and the New York Botanical Garden, the University of Antioquia has been carrying out
extensive field work in the Department of Antioquia. The data
base of the preliminary inventory work reveals nearly 6000
species, including fungi, algae, mosses and vascular plant
species. Many of these have been new discoveries (Atehortua
and Callejas, in preparation). Details of the other floristic
work in Colombia are given by Prance (1982).

Presently the National Herbaria Association is making an
effort to establish floristic regional inventories, attempting
to cover as much of the entire country as possible. In 1982,
the Institute of Natural Sciences at the National University,
began to publish Flora of Colombia, but most of the floristic
information is found in Flora Neotropic Monograph Series.

Development versus conservation is today the great dilemma
for developing countries. It is also a challenge for the
entire world, which depends upon the natural resources existing there. Solbrig (1990) points out that development without
conservation of natural resources is not sustainable, and
preservation of flora and fauna without development is uto-
pian. How are these countries going to solve this divergence?

It seems to be paradoxical that developed countries which
are causing the high environmental impacts with global implications are also imposing trade barriers on undeveloped countries, putting pressure on their economies and forcing them to
expand their agricultural frontiers.

According to ECLA (1983) and UNESCO (1981), during the

last 25 years, the land surface dedicated to agriculture and
grazing has increased by 728,890 sq. km. and this phenomenon
has come at the expense of former forest lands and savannas.
A World Bank report in 1978 estimated for all of Latin America, a conversion rate of natural forest of 50 to 10,000 sq.
km. per annum. For Colombia the average of forest destruction
has been estimate at 660,000 to 880,000 hectares per year
(Barco, 1990). This data does not include the large areas of
Amazonian forest devoted to uncontrolled gold mining activities that are creating a tremendous environmental impact in
Colombia as well as Brazil. The high revenues from the gold
mining seem to be a barrier for control by the local government.

One of the biggest problems of developing countries is the
increase of human population which leads to further alteration
of natural ecosystems and the expansion of food producing
agriculture. With the rapid human growth, needs for food and
commodities have increased proportionately. Unfortunately,
improvements mean more stress on natural resources, environment and quality of life. The expansion of the agricultural
frontiers means destruction of natural habitats and more
threat of extinction of our genetic resources.

The density of Colombia's population is approximately 30
inhabitants per sq. km., but there is a high degree of anal-
phabetism as well as poverty. Those socio-economic factors,
together with the fragility of the ecosystem, are leading this
country to unpredictable consequences, transforming this phytogeographic area from one of great megadiversity into one of
great megaextinction. Conservative calculations by IUCN,
cited by Andrade (1990) estimated that in Colombia, 1000 plant
species are under the threat of extinction, without taking
into account other organisms.

The National Renewable Natural Resources Institute
(INDERENA), the government agency created in 1975, has under
its protection the 29 national parks and the 4 Sanctuaries of
Fauna and Flora. However, this institution does not have adequate financial support, well-educated and trained experts;
therefore its function is not well-assumed.

Consequently, the panorama for *in situ* conservation does

not seem to be very optimistic and reliable. What are the
real alternatives for conservation of natural resources in
developing countries? There is no doubt that *in situ* conservation is the best for plant genetic resources, but how
reliable is this option for Latin American countries? How
could this goal be achieved without interfering with their
legitimate development?

In my view point, *in situ* conservation in Colombia is
unbridgeable with the socio-economical situation of today
unless some drastic socio-political changes take place. In
Colombia, political corruption, social disruption, poverty,
drug-related activities and guerrilla activities have generated a thoroughly chaotic and violent situation. Moreover, in
spite of the government good will to protect natural resources
and environmental impacts, little has been done to create a
formal and strong commitment to work a national goal that
guarantees the conservation of this patrimony for the present
and future of the entire society. Most of the difficulties
come from the increasing external debt, opening of international trades and improvement of living conditions for nearly
30% of the population; thus environmental protection is not
the first priority. Some attempts have been made. During the
presidency of Virgilio Barco, the Colombian government
returned to the Indian tribes approximately 18 million hectares, known as Indian reservation, as a major step for conservation of the lowland Amazonian forest.

On the other hand, non-governmental organizations, such as
Fundacio para la Educacion Superior (FES) in association with
WWF, in 1982 bought an area known as La Planada (southwest
Colombia in the Department of Marino) in order to establish a
natural reserve in this tropical ecosystem. La Fundacion Natura did something similar through an agreement with the Public
Water Service Reservoir of Bogota to preserve a natural Andean
forest known as Carpanta. Perhaps those places, if maintained
without human intervention, should be considered as good natural germplasm banks for *in situ* conservation of plant genetic
resources. However, someone should estimate the costs to
guarantee good maintenance and preservation, and to provide
resources for research projects and monitoring their species.

Being more aware of our socio-economic reality, *ex situ* conservation through a germplasm bank could be considered as an additional and reliable alternative to some future prospecting. What is the status of *ex situ* conservation of wild plants in Colombia? The major institutions supported by the government and the international agencies are CIAT and ICA, both of which engage in conservation programs of important crops and their wild relatives; for non-crop plants, there is no formal and well-financed institution dealing with *ex situ* preservation of wild plant species.

Colombia has 6 major botanical gardens, two of which are part of the University Agronomy Faculties. Also, there are some non-governmental organizations (ON_g) involved in conservation issues together with the IUCN World Conservation Strategy, such as the Fundacion Natura, Fundacion Pro-Sierra Nevada de Santa Marta and Herencia Verde. Along with those institutions, the main governmental research supporting agency is COLCIENCIAS and some other private foundations, such as FES and FEN, which are now giving some priorities to supporting research projects on environmental and conservation programs.

Although the botanical gardens could be good places for *ex situ* conservation of wild species and active research programs (Williams and Creech, 1987), they do not have enough funds to make well-trained scientists available to develop and maintain a computerized data base record system of the wild living collections and accessions, nor to carry out quality research projects on exploration, collection, banking, evaluation, documentation and breeding for enhancement. Therefore, for most of the Colombian botanical gardens, programs for *ex situ* conservation are beyond their scope and possibilities. [Details of the role played by the South American Botanical Gardens are given by Forero (1987).]

If Colombia wants to accomplish some, if not all, of the conservation world strategy goals, it would be necessary to reinforce those gardens already founded, instead of creating new ones, and build a National Germplasm System, together with a network of universities, scientific institutions, foundations and non-governmental organizations engaged in conservation and research programs. This network could be a prelimi-

nary data base to design a National Conservation Strategy and
to gather critical information for future development, with
the following aims:

1. What plant species are the priority for *ex situ* conservation?

2. What are the infrastructure and human resources needed
for a short, middle and long term prospecting?

3. How many scientists are already well-prepared to take
this unpostponable task and to achieve the goal?

4. What are the main research areas wherein it is necessary to concentrate our efforts?

5. What will the real cost be of *ex situ* conservation and
how should it be prospected with the time?

6. Who is going to fund this national and international
imperative?

7. How can an international network be built up with
institutions and foundations dealing with tropical plant conservation programs and how much can be learned from their
experience?

The first step toward the creation of a germplasm bank has
already been taken by the Joaquin Antonio Uribe Botanical Garden in Medellin, which has submitted a research project to
COLCIENCIAS, with the scientific cooperation of the two major
universities. The project is expected to provide funds for
specific training courses for professionals, undergraduate and
doctoral students.

In its first stage the project will begin with the domestication of the two wild species of the genus *Heliconia*. The
development of the first stage will provide the necessary
experience to undertake the conservation of more species of
larger and/or more critical groups. The garden is open to
receive cooperation of any institution that can provide funds,
experience and knowledge on germplasm bank techniques for
tropical plant species.

Finally, I would like to state that *in situ* as well as *ex
situ* conservation for development countries is an unpostponable challenge and task, a race against time. It is a global
imperative to secure human survival.

Literature Cited

Andrade, G. 1990. Colombia: Megadiversidad o Megaextinction. *Ecologica* 5: 4-10.

Barco, V. 1990. Eco-renta para el Tercer Mundo. A summary of the Colombian president speech held at the Royal Botanical Garden, Kew and Gala Foundation. *Ecologica* 5: 15-19.

Forero, E. 1987. 80,000 plants in South America: The case for creating more botanic gardens. Pp. 228-237. In: D. Bramwell, D. Hamann, V. Heywood and H. Synge (eds.), *Botanic gardens and world conservation strategy*. Academic Press, London.

Forero, E. and A. H. Gentry. 1989. Lista anotada de las plantas del Departmento del Choco, Colombia. Instituto de Ciencias Naturales, Museo de Historia Natural. Universidad Nacional de Colombia. Ed. Guadalupe Ltda. Bogota, D. E.

Gentry, A. 1982. Phytogeographic patterns as evidence for a Choco Refuge. In: G. T. Prance (ed.), *Biological diversification in the tropics*. Columbia University Press, New York.

Miller, D., A. Y. Rossman and J. H. Kirkbride. 1989. Systematics, diversity and germplasm. Pp. 3-11. In: L. Knutson and A. K. Stoner (eds.), [13] *Biotic diversity and germplasm preservation, global imperatives*. Klewer Academic Publishers, Boston.

Prance, G. T. 1977. Floristic inventory of the tropics: where do we stand? *Ann. Missouri Bot. Gard.* 64(4): 659-684.

_____ (ed.). 1982. *Biological diversification in the tropics*. Columbia University Press, New York.

Raven, P. H. 1987. The scope of the plant conservation problem world-wide. Pp. 19-29. In: D. Bramwell, D. Hamann, V. Heywood and H. Synge (eds.), *Botanic gardens and world conservation strategy*. Academic Press, London.

Solbrig, O. T. 1990. Conservation and development in tropical South America. *Harvard Papers in Bot.* 2: 1-10.

Williams, T. J. and J. L. Creech. 1987. Genetic conservation and the role of botanic gardens. Pp. 161-173. In: D. Bramwell, D. Hamann, V. Heywood and H. Synge (eds.),

Botanic gardens and world conservation strategy.   Academic
Press, London.

THE ECONOMIC IMPACT OF NOVEL GENES IN PLANT BIOTECHNOLOGY: NOT WITHOUT STRONG INTELLECTUAL PROPERTY RIGHTS

Steven C. Price

Office of Intellectual Property, Iowa State University
Ames, Iowa 50010 U.S.A.

Summary - The historical basis of patents is reviewed, along with the development of intellectual property rights as they pertain to plants and biotechnology. The thesis is advanced that patents and licensing will not inhibit research and development, but actually promote research development amid the free exchange of information and materials. This is a world-wide concept that applies equally to developed and developing countries.

INTRODUCTION

Since 1787, when the Kew Royal Botanical Gardens appointed Mr. David Nelson to accompany Captain Bligh on the ill-fated Bounty to collect and transport the breadfruit tree from Tahiti to the West Indies (Juma, 1989), to the present, there have been problems surrounding the global movement of germplasm. It will be my attempt to show how we will continue to have such problems unless we re-orient our attitudes, methods, and policies concerning global intellectual property protection. Intellectual property protection has emerged as a topic as relevant as traditional scientific questions concerning the handling of germplasm.

Germplasm sources were classified by Harlan and DeWet in 1971 as ranging from the "primary gene pool", which involve reservoirs within the traditional concept of a biological species, to the tertiary gene pool, which include species which were crossable but produce hybrids that are "anomalous, lethal, or completely sterile." The tertiary gene pool was considered by him to represent the "extreme outer limit of the potential gene pool of a crop." (Harlan, 1975).

Until recently, plant breeders who were desirous of introducing genetic variability into crop species had to use either

genetically divergent germplasm, or various forms of mutagen-
sis. Although genetic engineering was only first revealed in
enabling publications in 1973 (Jackson et al., 1972; Cohen et
al., 1973) and first convincingly demonstrated in plants in
1984 (Horsh et al., 1984; De Block et al., 1984), it is
already providing to plant breeders accessibility to gene
pools well outside the generic, familial, or even kingdom lev-
el-taxonomic boundaries. We now have numerous examples of
potentially economically important genes, which I will discuss
shortly, that were first identified in bacteria, and have not
only been moved into higher plants, but will be ready for commercialization in the early 1990's. Therefore, the entire
biosphere is indeed not too large of a realm to consider when
canvasing for new sources of germplasm, and indeed we only
seem limited by our imaginations. At a conference held in
1985 titled "Crop Productivity--Research Imperatives Revisited" (Gibbs and Carlson, 1985) various scientific and policy
issues were identified which were seen as world needs for the
coming decades in agriculture. In the arena of public policy,
it was advised that: 1) there be an increased interaction
between public and private sectors; 2) there be an increased
collaboration between developing and developed countries; and
3) there should be a sharing and protection of genetic
resources.

Similarly, it was recommended at a 1988 conference commis-
sioned by the USDA, titled: "An Evaluation of the Issues,
Challenges, and Opportunities Related to Plant Patenting"
that "...it is important to support better characterization of
material in gene banks so that characterization is known and
in the public domain, available, and therefore, not subject to
IPR" (intellectual property rights).

At a conference held in 1989 sponsored by USAID (Agency
for International Development), which was concerned with
establishing priorities in plant biotechnology that would
benefit developing countries, it was recommended that intellectual property rights should be developed which would
"...make proprietary (technology)...available to developing
countries in a timely manner."(Anonymous, 1990a).

It is these latter points--1) opposition to intellectual

property protection, and 2) a recommendation to enhance the
sharing of genetic resources, that I wish to address. It is,
in my opinion, too often assumed that intellectual property
protection will result in a decreased sharing of germplasm and
a stifling of research and innovation.

In addition to asking scientific questions relative to the
handling of germplasm, it may also be appropriate to ask if we
are managing germplasm in such a way as to maximize its use
economically. It is, after all, the economic use in which we
are ultimately most interested; the public will only be benefited if companies incorporate the germplasm into cultivars
which can then be marketed, distributed, grown, harvested,
refined, and placed into a product the consumer can use.

It is my intent in this paper to advance the thesis that
proper intellectual property rights must be among the first
steps to ensure maximal economic use. Indeed, to either
intentionally or unintentionally **not** allow for these activities may have the consequence that the public will not be
fully benefited. Finally, intellectual property protection
should have the seemingly paradoxical effect of **enhancing**
germplasm exchange.

BACKGROUND

Intellectual property rights are, simply, "An aggregate of
rights resulting from the creative efforts of the mind,"
and the term is used for data, know how, utility patents,
plant patents, plant variety certificates, plant breeder's
rights, trade secrets, etc.

It is generally recognized, however, that the strongest
form of intellectual property protection that can be offered
is that of utility patent.

What is a patent? A patent is the legal right that permits the owner to exclude others from making, using, and selling an invention. Exclusionary right lasts for 17 years
after the patent is issued. In return, the inventor must disclose his or her invention to the world in such detail that
other people will be able to make and use it. A utility
patent is essentially a legal right that permits the owner to
exclude others from making, using, and selling an invention.

The exclusionary right lasts for 17 years in the U.S., and for
a similar period in other parts of the world, after the patent
is issued. A utility patent has four essential elements (From
Patent Statute, Title 35 U.S. Code, Sections 101, 102, 103):
it must cover certain subject matter; it must be useful; it
must be novel, and it must be an unobvious extension of what
already is known to the public. It has two notable features:
1) the patent enables an inventor to teach the world (it is
not a secret); and 2) it gives a monopoly on its use to the
inventor (more accurately, the title holder). In return for
this monopoly, the inventor must disclose his or her invention
to the world in such detail that other people will be able to
make and use it. A major world-wide debate has surrounded
what constitutes patentable subject matter.

Intellectual property protection for plants and products
of biotechnology have resulted in what today is truly a patch-
work of world-wide intellectual property laws. In Europe,
this "grainy environment" has long been recognized, and 13
member states of the European Common Market signed the Convention on the Grant of European Patents in 1973 in an attempt to
"harmonize" various European patent laws (Table 1).

Table 1 European Patent Convention

•Grants European Utility Patents
•Provides patents for inventions that are useful, new, and
    involve an inventive step
•Excludes plant and animal varieties and "biological pro-
    cesses" that are conventional
•Will grant patents for gene transfer technologies or for the
    gene itself

However, what constitutes patentable subject matter has
been of considerable concern, since plant and animal varieties
are specifically excluded from a European Patent, though
microbes, DNA molecules or subcellular units such as proteins,
and plasmids are allowed. Also, "...essentially biological
processes for the production of plants or animals..." were

excluded (Office of Technology Assessment, 1990).

Plant varieties were excluded from the European Patent Convention because such "Plant Breeder Rights" were already provided under the International Union for the Protection of New Varieties of Plants (UPOV), which had its origins in the 1950's, was signed in 1961, and revised in 1978 by 17 countries (Belgium, Denmark, France, Germany, Hungary, Ireland, Israel, Italy, Japan, Netherlands, New Zealand, South Africa, Spain, Sweden, Switzerland, United Kingdom, and the United States). Such Plant Breeder Rights give limited patent-like protection (Table 2).

Table 2. International Union for the Protection of New Varieties of Plants (UPOV)

•Provides protection for plant varieties
•Both sexually and vegetatively propagated plants
•Plants must be distinct and novel
•Most countries permit development of new varieties, and if derivatives, would not infringe
•Some countries have mandatory licensing

In addition, some individual countries (Germany, France, and Australia) allow national utility patent protection for plant varieties not covered by plant variety protection. However, utility patent protection is not available in Argentina, Brazil, Canada, China, India, the Philippines, and the USSR (Jondle, 1989). Apparently Africa, excluding South Africa and Kenya, has no form of protection for plants.

Regardless, in Europe, the recognition that the European Patent Convention, individual country patent statutes, and UPOV, do not offer adequate protection for biotechnological inventions has been a serious source of concern for the European Commission. Therefore, on the 17th of October, 1988, the Commission submitted a Directive that accepts existing conventions, buts seeks to provide legal interpretations that are more favorable to patentees of biotechnological inventions. The time frame was that these Directives should have been enacted by 1990, but my understanding is that of April, 1991, they still have not been. As Duesing and Raeber (1989) point out, there are serious limitations associated with the Directive. For example, under the Directive, plants and animals that are produced by known biotechnological processes would not be patentable. Furthermore, three years, after a patent is granted, if another inventor develops a variety similar to the first patented, the second would be required to be issued a compulsory nonexclusive license.

In the United States, patent protection for new and novel asexually produced plants has been available since 1930 with the passage of the Plant Patent Act (Table 3).

Table 3. U. S. Plant Patents (U.S.C. Title 35, Section 161).

•Provides for the patenting of asexually produced plants--
   uncultivated and tuber propagated plants are excluded

•Novelty, distinct, non-obvious requirements, as with Section
   101 patents

•Single claim allowed to the entire plant

This act allows for patents with one claim covering the whole plant, and coverage lasts for 17 years. Since 1930, over 6000 plant patents have been issued (Jondle, 1989). In the U.S. it was recognized at least 25 years ago that plant breeders of sexually reproduced crops needed patent-like protection, since sexually reproduced plants at that time were excluded from utility patents, and hence the Plant Variety Protection Act of 1970 was enacted (Table 4).

Table 4. U. S. Plant Variety Protection Act of 1970.

•Gives a Plant Variety Certificate to plant varieties
•Covers sexually reproduced plants other than fungi, bacteria,
    or hybrids, including seeds and transplants
•Gives right to exclude others from selling, reproducing,
    importing, exporting, or using it to produce a hybrid in
    U.S. for period of 18 years
•Does not exclude use for research, plant breeding, or for
    developing a new variety or hybrid--subject of judicial
    interpretation
•Farmer exemption--they can save seed and offer for sale to
    another farmer, with some limitations--subject of judicial
    interpretation

This act gives patent-like protection for 18 years to holders
of a Plant Variety Protection Certificate (PVP) (Table 5).

Table 5. U.S. Requirements for Obtaining a Plant Variety
    Certificate.

•Must be sexually reproduced

•Must be novel: distinct, uniform, and stable

•Novelty identified by an "Application"

•No claims are required

•Deposit of 2,500 seeds in public depository

    However, there are at least two notable problems associ-
ated plant variety protection by PVP, and have been subject to
judicial interpretation: 1) the farmer exemption which allows
farmers that produce up to 49% of their crop with the protected variety, and sell it without compensation back to the

owner of the protected variety; and 2) the research exemption
which allows for any derivatives of the protected variety, no
matter how cosmetic or trivial, itself to be protected under
the PVP law, also without compensation back to the title
holder. Since 1970, over 2100 PVP Certificates have been
issued (Jondle, 1989).

During the last decade, debate in the United States has
surrounded the determination of what constitutes patentable
subject matter for utility patents, since utility patent protection was not allowed for life forms. A series of benchmark
decisions, starting with the "Chakrabarty" case in 1980 that
dealt with conventionally altered bacteria, and ending with
the "Hibberd" (1985) case which involved tryptophan overproducing maize plants, have resulted in rulings that plants are
now patentable subject matter under the utility patent system
(Table 6).

RATIONALE FOR INTELLECTUAL PROPERTY PROTECTION

The existence of utility patents, Plant Variety Certificates, and plant patents gives modern plant scientists an
acceptable range of intellectual property protection in the
United States for a vast range of genes, proteins, plant
parts, seeds, tissues, and varieties. They assure breeders
and researchers that they will have intellectual property protection homologous to that enjoyed by other inventors, and
provides industry an assurance that they will be able to
recover costs by benefiting exclusively from the commercialization of their products. However, such is not the case in
other parts of the world. In many developing countries,
intellectual property protection and enforcement are so defi-
cient as to assure that these countries will not be benefited
economically to the fullest extent by the new technologies.

The extent to which intellectual property protection has
in the past served to spur economic development and research
can be approximated by several statistics. Britain's private
breeding industry was apparently "saved" by the adoption of a
breeders rights system in 1964. In the U.S., by 1980 "three
to six times more new varieties of wheat, soybeans, and cotton
were produced "...after passage of the 1970 act..." (Barton,

Table 6. U.S. History of What Constitutes Patentable Subject Matter.

U.S. Patent No. 141,072-To Louis Pasteur in 1873 a claim of "Yeast, free from organic germs of disease..."

Plant Patent Law of 1930-Included new and distinct asexually reproduced plants, including cultivated sports, mutants, hybrids, and seedlings. Excluded tuber producing plants and bacteria.

Diamond v. Chakrabarty case-447 US 303 (1980)-Landmark ruling that a live, human-made microorganism is patentable subject matter under 35 USC 101 as a composition of matter.

U.S. Patent No. 4,259,444 (1981)--Resulting Chakrabarty general patent covering man-made microorganisms that have oil-degrading plasmids

U.S. Patent No. 4,642,411 (1987)--"Hibberd" patent covering tryptophan overproducing mutants of cereal crops. Covered tissue cultures, seeds, plants, and hybrids.

Ex parte Allen, 2 USPQ 2D 1425.(1987)--The Patent and Trademark Office Board of Appeals ruled that animals constitute patentable subject matter. Based on polyploid oysters. Did not involve genetic engineering.

1988. U.S. Patent on genetically engineered mouse.

1982). The number of private breeding programs in soybeans alone has increased from fewer than 5 to more than 30 since the PVP act (Anonymous, 1989). Since the allowance of plants as patentable subject matter in the United States, the number of patent applications for utility patents has gone from 73 applications in fiscal year 1986, to perhaps more than 400 in

1989 (Warren, 1989).

It is not generally appreciated that some form of monopol-
istic control has been found to be necessary to propel western
economic development for the last 2000 years: the progression
of industrial society has coevolved with the development of
patent systems. Today's patent system has its origins as far
back as 300 B.C., where confectioners or cooks who invented
"any peculiar and excellent dish" were given exclusive rights
to make it for one year "in order that others might be induced
to labour at excelling in such pursuits." In the Middle Ages,
Kings granted monopolies, in the form of "letters of patents,"
to merchants giving exclusive right to sell commodities and
inventions, and was later modified to cover only inventions.
The origin of the patent system in use today stems, with few
changes, from a statute from the republic of Venice, March 19,
1474. From Venice the system spread to Germany, and reached
England in 1561. In the United States, the Constitution
authorized Congress to enact laws to "promote the progress of
science and useful arts, by securing for limited times to
authors and inventors the exclusive rights to their writings
and discoveries." In 1790, the first patent statute was
enacted by Congress. It has very few differences from the
Venice statute, now 514 years old. Both required novelty,
utility, and gives rights of ownership for a limited period of
time. The U.S. Constitution: Article 1, Section 8 authorized
Congress to enact laws to "promote the progress of science and
useful arts, by securing for limited times to authors and
inventors the exclusive rights to their writings and discoveries."

Today, there is great concern in the United States for the
pirating of its technology (Hanson, 1988). A report from the
International Trade Commission estimates that U.S. companies
lost $61 billion in 1986 because of poor intellectual property
protection, across all technologies. Fourteen countries have
been identified as having impacts of more than $100 million
each, in descending order: Taiwan, Mexico, South Korea, Brazil, China, Canada, India, Japan, Nigeria, Hong Kong, Saudi
Arabia, Indonesia, Italy, and Spain. The United States, the
European Economic Community, and Japan claim that $3 billion

per year is lost by the pirating of their seeds and drugs (Anonymous, 1990b).

We can be sure that companies' and universities' sensitivities to their inventions in biotechnology will be reflected in their unwillingness to trade technology with those countries that have poor or nonexistent intellectual property laws. This alone will insure limited access by these countries and others that are perceived to have poor patent protection and enforcement. The importance of intellectual property protection to developed countries can be seen by the centrality of intellectual property rights to the GATT (General Agreement on Tariffs and Trade) negotiations, of which plant intellectual property is of concern.

INTELLECTUAL PROPERTY PROTECTION OF ECONOMICALLY USEFUL GENES

A measure of how important intellectual property protection is to institutions that are involved with plant biotechnology can be ascertained by a survey of the IP status of economically useful genes in this new, fast moving field. Table 7 covers a wide array of economically important genes in plant biotechnology, and reflect the majority of publicly known plant biotechnology efforts. We can see that the majority of the listed genes are not only associated with companies, but have some form of intellectual property protection associated with them: patent protection gives institutions the necessary incentive to incur the costs associated with moving these traits into the public domain, and the assurance that if the patent is infringed, there will be legal recourse. To not allow companies to have some form of a monopoly will probably result in the technology either not being developed, or significantly delayed. In both instances, of course, the public is not benefited.

Therefore, to advocate no IP protection for germplasm collections, as advocated by the USDA 1988 conference (Anonymous, 1989) will most assuredly insure that economically valuable genes will not be used by the companies that could otherwise take them to the market in an efficient and timely manner.

Guaranteeing access to germplasm emerges as a primary con-

Table 7. Important genes in plant biotechnology ownership.
*P = Proprietary protection; NP = no proprietary protection;
blank = unknown

Herbicide Resistance

Source	Recipient	Active Ingredient	Institution*
?	?	Metribuzin	Calgene[P] Mobay
Klebsiella	Tobacco Tomato	Bromoxynil	Calgene[P]
Nicotiana	?	Chlorosulfuron	Dupont[P]
Nicotiana	?	Sulfometuron	Dupont[P]
Salmonella *Streptomyces*	Tobacco Rape	Glyphosate	Calgene[P]
Brassica campestrus	Rape	Atrazine	Univ. of Guelph[P]
Petunia	Cotton Soybeans	Glyphosate	Monsanto[P]
	Canola	Atrazine	Ciba-Geigy
Streptomyces	Potato Tobacco Tomato	Phosphinotricin	Plant Genetic Systems[P]
Zea	Corn	Imidazolinones	•American Cyanamide •Pioneer Hi-Bred[P] •Molecular Genetics
Alcaligenes	Tobacco	2,4-D	Washington State University[P]
Allelopathic Compounds			Many

Table 7.   (Continued)
Important genes in plant biotechnology ownership.   *P = Proprietary protection; NP = no proprietary protection; blank =
unknown

Insect Resistance

Source	Recipient	"Insecticide"	Institution*
Bacillus thuringiensis	*Clavibacter >* Corn	InCide vaccine	Crop Genetics International[P]
Bacillus thuringiensis	Cotton	BT control protein	Monsanto[P]
Bacillus thuringiensis	Cotton, Dry Beans, Poplars, Cranberries	BT control protein	Agracetus[P]
Bacillus thuringiensis	Cotton	BT control protein	Calgene
Bacillus thuringiensis	*Pseudomonas*	BT control protein	Mycogen[P]
Rice stripe virus	Rice	Rice stripe virus coat protein	Japan Ministry of Agriculture, Forestry, and Fisheries[NP]
Potato Virus X and Y (PVX and PVY)	Potato Tobacco	PVX and PVY coat protein	•Monsanto •The Rockefeller University
Potato Virus X (PVX)	Potato	PVX coat protein	Mogen International[P]
Tobacco Mosaic Virus (TMV)	Tomato	TMV coat protein	Monsanto
Alfalfa Mosaic Virus (AIMV)	Tobacco Tomato	AIMV coat protein	•Monsanto •Washington University
Alfalfa Mosaic Virus (AIMV)	Tobacco	AIMV coat protein	Agrigenetics

Table 7. (Continued)
Important genes in plant biotechnology ownership. *P = Proprietary protection; NP = no proprietary protection; blank =
unknown

Insect Resistance

Alfalfa Mosaic Virus (AIMV)	Tobacco	AIMV coat protein	State University of Leiden
Tobacco Mosaic Virus (TMV)	Tobacco tomato	TMV coat protein	•Monsanto •Washington University
Cucumber Mosaic Virus (CMV)	Tobacco	CMV coat protein	•Monsanto •The Rockefeller University
Tobacco Streak Virus (TSV)	Tobacco	TSV coat protein	State University of Leiden
Tobacco Rattle Virus (TRV)	Tobacco	TRV coat protein	State University of Leiden
Soybean Mosaic Virus (SMV)	Tobacco	SMV coat protein	Monsanto
Tobacco	Tobacco	Proteinase Inhibitor II	Iowa State University[P]
Potatoes Tomatoes	Tobacco	Proteinase Inhibitor II	Washington State University
Tobacco Ringspot Virus	?	Antisense RNA	CSIRO
Cowpea		Cowpea Trypsin Inhibitor	Agricultural Genetics Company[P]
		Chitinase	DNA Plant Technology[P]

Table 7.   (Continued)

Important genes in plant biotechnology ownership.   *P = Proprietary protection; NP = no proprietary protection; blank = unknown

Nitrogen Fixation

Source	Recipient	Institution*
Rhizobium	Alfalfa	Biotechnica Agriculture

Molecular Farming

Gene	Host	Institution*
2S Albumins linked to Enkephalins	*Arabidopsis*	•Plant Genetic Systems[P] •University of Ghent
Human Serum Albumin	Potato Tomato	Mogen International[P]
Geneware[TM] Melanin	Tobacco cell culture	Biosource Genetics Corp.
Polyhydroxybutyrate	*Alcalignenes>* "plants"	•Pioneer-Hibred •Massachusetts Institute of Technology[P]
Pokeweed antiviral protein from *Phytolacca*	*E. Coli*	Agricultural Genetics Company[NP]
Pharmaceuticals	"Plants"	DNA Plant Technology
"human interest proteins"	"Plants"	•Phytodynamics •British Petroleum

Table 7. (Continued)
Important genes in plant biotechnology ownership. *P = Proprietary protection; NP = no proprietary protection; blank =
unknown

Food Quality

Gene/Phenotype	Host	Trait	Institution[*]
Antisense polygalacturonase	Tomato	Reduced softening	•ICI Seeds[P] •University of Nottingham
Antisense polygalacturonase	Tomato	Reduced softening	Calgene[P]
Acetyl Coenzyme A Carboxylase	Carrot	Fatty acid modification	Iowa State University[P]
High Oleic Acid	Canola	High Oleic fatty acids	University of Idaho
High Amylose Starch	Corn	High Amylose Starch	Iowa State University[P]
Low Linolenic Acid	Soybeans	Low Linolenic fatty acids	Iowa State University[P]
Low Palmitic Acid	Soybeans	Low Palmitic (saturated) fatty acids	Iowa State University[P]

Modifier Genes

Gene	Institution[*]
Small Subunit Promotor 301 from Petunia	DNA Plant Technology[P]

Heat Shock Proteins

Gene	Institution[*]
Heat Shock Producing gene from yeast	•Iowa State University[P] •Yale University[P]

Table 7. (Continued)

Important genes in plant biotechnology ownership. *P = Proprietary protection; NP = no proprietary protection; blank = unknown

Gene Maps

Type	Plant	Institutions*
Whole Genome	Many	USDA Others
RFLP	Wheat Barley	•Agricultural and Food Research Council's Institute of Plant Science Research and: •Agricultural Genetics •Nickerson International Seed •ICI •CIBA-Geigy •Plant Breeding International
RFLP	Corn	Pioneer Hi-Bred[P]
RFLP	Corn	ICI Garst[P]
RFLP	Soybeans	Dupont[P]
RFLP	Corn Tomato	•NPI[P] •Dekalb Pfizer •H.J. Heinz

cern in almost all policy discussions pertaining to developing countries. Oftentimes, intellectual property protection has been identified at distinguished and influential conferences as a practice that will result in an <u>inhibition</u> of germplasm exchange (Anonymous, 1989). The reality of the market place, however, is that companies and universities will keep to the

IP trajectory they have had in the past--the research and
development costs and the economic rewards are too high to
alter these policies.

Evidence of the importance of IP to modern industrialized
societies can be seen by the use of IP as leverage in trade
negotiations, such as GATT, and by the ever increasing power
of IP protection and its adoption by other countries. For
example, Canada has finally adopted a Plant Breeder's Act, and
will become a signatory to UPOV (Anonymous, 1990b). The American Seed Trade Association continues to move towards a position advocating strong intellectual property protection and
enforcement, contrary to earlier positions (ASTA, 1988, 1990).

The uncertainty that IP itself will result in decreased
germplasm exchange between developed countries will probably
have less to do with IP than with the competitiveness of modern day research. Today, there is substantial sharing of
information and materials between competitors. However, when
IP is involved, there are appropriate instruments to guarantee
economic protection for all parties, and these are being used
more and more by modern research laboratories (Table 8).

Prior to the filing of patent applications, <u>confiden-
tiality agreements</u> can be executed, simply acknowledging that
the information to be shared is confidential and is not to be
used for economic purposes. The sharing of seeds themselves
can be preceded with <u>research agreements</u> that acknowledge that
the seed is to be used for research purposes, and that intel-

---

Table 8. Legal instruments for facilitating information and
material exchange.

•Confidentiality Agreements
•Research Agreements
•Contracts
•License Agreements

---

lectual property rights reside with the originator. For
example, once patent applications are filed, then the content
of the patent application can be shared with other researchers

with little concern for losing one's competitive advantage, since the date of filing consisted a legal priority date in most foreign countries. The contents of the patent application itself is published in most countries after 18 months, prior to the issuance of a patent.

Contracts can specify the terms and the conditions for the transference of rights in intellectual property. The transference of rights to practice under a patent are all dictated by a license agreement (Table 9).

Table 9. License agreements.

Advantages: Overcome many structural problems or philosophical objections inherent in other forms of intellectual property protection.

Examples:
- Ability to conduct research
- Access to germplasm
- Geographical limitations
- Term of agreement
- Third party/vendor requirements
- Local planting politics
- Royalty distributions

License agreements can constitute an extremely powerful instrument, where many concerns, political and philosophical, can be accommodated. In fact, they can be considered a "keystone" for holding together existing laws and diverse views. For example: license agreements specify who owns the technology, the degree, if any, of economic remuneration to the licensor, ownership of future research results, commercialization requirements, and germplasm exchange. Specifically, license agreements provide for:

A. *Ability to conduct research*. There is some question as to whether or not utility patents extend the concept of "make, use or sell" to include excluding others from doing research. This is of course of critical concern to universities, whose primary mission is to promote research. It turns out that if you ask an attorney if a utility patent can be used to exclude others from doing research, what you get is an uncomfortable answer.   It is my understanding that a doctrine has evolved over the years (called the "experimental purpose doctrine") whereby an accused infringer's making or using what is patented may be excused if it is solely for an experimental or other non-profit purpose.   Therefore, if research has a commercial end, then it could be an infringement.   If this is a concern in a specific situation, then the nature of the experimental use can be handled by a license agreement, wherein such use is either allowed or not allowed.

B. *Access to germplasm*.   Another concern is that if germplasm is patented, then it will not be accessible to the general public.   This can also be specified in a license agreement.   At Iowa State University, we have recently signed a license agreement whereby we required the licensee to make seed available to all who want it--with a few restrictions, such as although universities get seed for free, they must agree it is for research purposes only, and if they develop commercial material, then they must commercialize it through our licensee.

C. *Geographical limitations*. It is important to realize that if one has filed patents world-wide, the license agreement can be used to specify where in the world the licensee(s) can practice.   For example, if a patent has issued in both the United States and in Europe, one can still restrict a licensee to specific regions of the world.   Furthermore, one can specify that royalty is to be collected in countries which do not have patent coverage.

D. *Term of the agreement*.   Another important item is how long do you want an agreement in effect?   It is important to point out that even though a patent may only good for 17 years, one can still collect royalty for an indefinite period of time, especially if some "know-how" is passed along with the patent rights.

E. *Local planting politics*.   Pressure may be exerted from local constituencies to restrict commercial production to specific regions; license agreements again can specify this kind of item.

F. *Royalty distributions*. One sometimes hears the complaint that they have an objection to patenting or PVP because they are opposed to royalty collection. However, this has nothing to do with the patent or PVP *per se*, but in any event the amount and distribution of royalties can be specified in the agreement--how much to the inventors, how much for research, etc.

All of these legal documents have become, or are becoming, commonplace tools among researchers. These "permits" are indeed burdensome, and require more paperwork, but they reflect the reality and needs of the market place that we all find ourselves in. Developing countries must become accustomed to these tools, incorporate them into their everyday tool-kits and be as comfortable with them as they are comfortable with the various techniques of modern plant biotechnology. There is evidence that the international community is realizing the importance of strong intellectual property protection for developing nations. The Keystone International Dialogue in January of 1990 concluded that "...the evaluation, use and maintenance of plants may be encouraged indirectly by the patent system..." (Anonymous, 1990b). The development of plant breeding and the rapid progression the new biotechnology has led breeders, biotechnologists, and investors to demand that their research investments be rewarded. To accommodate this will insure that all of us will be rewarded.

Literature Cited

Anonymous. 1989. Workshop Summary Report: Relevant forms of legal protection. Pp.175-190. <u>In</u>: B. E. Caldwell and J. A. Schillinger, (eds.), *Intellectual property rights associated with plants*. Crop Science Society.

Anonymous. 1990a. *Plant biotechnology research for developing countries*. National Academy Press, Washington D.C.

Anonymous. 1990b. Uruguay round of GATT provides new forum for debating germplasm ownership issues. *Diversity* 6: 3-4.

ASTA. 1988. Statement approved by the Executive Committee of the American Seed Trade Association, January.

ASTA. 1990. Statement approved by the Executive Committee of the American Seed Trade Association, June.

Barton, J. H. 1982. The international breeder's rights system and crop plant innovation. *Science* 216: 1071-1075

Cohen, S., A. Chang, H. Boyer, and R. Helling. 1973. Construction of biologically functional bacterial plasmids *in vitro*. *Proc. Nat. Acad. Sci. 70(11): 3240-3244.*

De Block, M., L. Herrera-Estrella, M. Van Montagu, J. Schell and P. Zambryski. 1984. Expression of foreign genes in regenerated plants and in their progeny. *The EMBO Journal* 3(8): 1681-1689.

Duesing, J. H. and J. G. Raeber. 1989. Requirements of Industrial Firms for Intellectual Property Protection in Plant Breeding and Biotechnology. Presented at the Symposium: Proprietary rights for new plant material at the XII. EUCARPIA Congress, SCIENCE FOR PLANT BREEDING, held 27 Feb.-4 March

Gibbs, M. and C. Carlson, (eds.). 1985. Crop Productivity: Research Imperatives Revisited, an international conference held at Boyne Highlands Inn, October 13-18, 1985 and Airlie House, December 11-13, 1985.

Hanson, D. 1988. High cost to U.S. firms of lax intellectual property laws detailed. *Chemical and Engineering News*, March 14: 19-20.

Harlan, J. R. 1975. Our vanishing genetic resources. *Science* 188: 618-621.

_____ and J. M. J. De Wet. 1971. Toward a rational classification of cultivated plants. *Taxon* 20: 509-517.

Horsch, R., R. T. Fraley, S. G. Rogers, P. R. Sanders, A. Lloyd and N. Hoffman. 1984. Inheritance of functional foreign genes in plants. *Science* 223: 496-498.

Jackson, D., R. Symons and P. Berg. 1972. Biochemical method for inserting new genetic information into DNA of simian virus 40: circular SV40 DNA molecules containing lambda phage genes and the galactose operon of *E. coli*. *Proc. Nat. Acad. Sci. 69: 2904-2909.*

Jondle, R. J. 1989. Overview and status of plant proprietary rights. Pp. 5-15. In: B. E. Caldwell and J. A. Schillinger, (eds.) *Intellectual property rights associated with plants*. Crop Science Society.

Juma, C. 1989. *The gene hunters: Biotechnology and the scramble for seeds.* Princeton University Press.

Office of Technology Assessment. 1990. *New developments in biotechnology: Patenting of life.* Marcel Dekker, Inc.

Warren, C. F. 1989. Issues and Challenges in the administration of the patent law with regard to plants by the Patent and Trademark Office. Pp. 145-156. In: B. E. Caldwell and J. A. Schillinger (eds.), *Intellectual property rights associated with plants.* Crop Science Society.

GERMPLASM CONSERVATION AND ECONOMIC DEVELOPMENT

Luiz Antonio Barreto de Castro

CENARGEN-CP 102372
Brasilia DF 70770 Brazil

Summary - In July of 1992 representatives of most countries in the world will attend in the city of Rio de Janeiro,
Brazil, the UNCED - United Nations Conference on Environment
and Development. The agenda for this conference will certainly include controversial issues which have been discussed
for decades by developing and developed countries, without a
foreseeable consensus. These are, however, the issues of this
decade and can be no longer postponed. They include matters
such as environment protection, germplasm conservation and
availability, biological safety, intellectual property rights,
technological and economical development. At the root of all
the discussions are expectations among developed countries
that a significant effort is necessary on the part of developing countries that will lead to a common understanding and
strategy to approach the first four items. Developing countries view the need to reconcile this effort with goals which
ultimately lead to social and economical development. As a
scientist and as a member of the Working Party on Biotechnology for the UNCED, I have discussed these matters on several
occasions at international meetings. Our observations indicate that the interfaces among these issues are so significant
that no isolated solutions will be possible. I will present
my view and a very brief update on some of these matters, from
the perspective of a scientist working in a developing country
(Brazil) of Latin America.

INTELLECTUAL PROPERTY RIGHTS

Biotechnology in the 1990's has become a very powerful
tool, capable of creating male sterile plants by genetic engineering (Mariani *et al.*, 1990). Genetic engineering produced
in the last four years more than 1000 plants (if one considers
only the permits required to the USDA) which are either being

tested under confinement or being released to the environment
for extensive field testing. This was possible because the
so-called enabling technology was shared, after negotiation,
internationally among several private and public institutions
in developed countries. Scientists isolate and patent genes
which are cloned in expression vectors and license these products or processes to others for a price. A significant number
of developing countries have been mostly excluded from this
process due to positions contrary to the recognition of intellectual property rights in biotechnology, including plants and
living matters. This subject, which has met considerable
resistance on the part of developing countries during decades,
is being reviewed in several countries in Latin America. Last
year only two international workshops were conducted in Central and South America on the matter, including an impressive
participation of country representatives, and international
institutions dealing with the subject. The unofficial conclusion of these meetings indicate that these countries are at
least in favor of adopting a mechanism of recognition of Plant
Breeder's Right, like the UPOV convention which applies to
plant varieties. As for the patent of biotechnological products and processes, the tendency is to accept a patent as
recognized by the Paris Convention which distinguishes between
inventions which are patentable and discoveries which are not.

ENVIRONMENT, BIOTECHNOLOGY AND DEVELOPMENT

The world has never been so close to an understanding on
these issues which were banished from discussion by developed
and developing countries a couple of decades ago. The concept
of intellectual property protection is being, as mentioned
before, gradually accepted by developing countries. Equally
the concept that genetic resources constitute a common heritage of mankind demands conservation, which can only result
from a concerted international effort. It is, however,
increasingly clear by developed countries that environment and
biodiversity conservation has a cost, and that the access to
genetic resources available in developing countries has a cost
as much as the access to technology protected intellectually
has a price.

As proposed by an UNEP report (1990): "It is possible to envisage the payment of royalties when a product or process is derived from the use of protected biological resources". It is also becoming evident that the major issue of next century, among all cited, is germplasm conservation. Technologies such as RFLP (Botstein *et al.*, 1980) and more recently mapping of genomes using random primers (Williams *et al.*, 1990), associated with T-DNA insertional mutagenesis (Feldmann *et al.*, 1989) will reveal the genetic make-up of hybrids and more importantly, will relate genes to specific functions. Genes will then be one of the most important "commodities" of the next century.

As gene functions are gradually known and gene banks, such as proposed in this meeting are established, access to genetic resources and to costly plant breeding will be facilitated. The task will be to slow genetic erosion to the benefit of mankind. The establishment of an international fund to finance the protection of biological diversity has been proposed (UNEP, *op. cit.*), and is therefore vital. So the challenge before us, perhaps at the UNCED in 1992 is to elaborate a proposal which will use biotechnology to reach development, based upon an adequate mechanism of germplasm conservation and utilization and conditions which are biologically safe to society and to the environment.

AGENDA 21

One of the tasks of the Working Party on Biotechnology of the UNCED is to assist the preparatory committee in preparing the work program for biotechnology, and in particular, to respond to the request by the committee to develop the Agenda 21, envisaged to be an agreed program of action for the international community for the period beyond the conference and into the 21st century. An integrated program on biotechnology will require international cooperation to be achieved by extensive negotiation and agreement on the issues presented above.

The implementation of such a program will have to consider two essential components: a clear definition of priorities needed to meet a typical "market-oriented" tendency of biot-

echnology (Burrill *et al.*, 1990) and the magnitude of the
financial investments needed to produce the expected developmental impacts under safe procedures of environment conservation. In this respect, the intended program is distinct from
other programs, financed by international public funds which
provide "seed money" for the concerted activities of laboratories in different countries. Although these activities are
very important and needed, they often lead to results of academic and scientific importance, which, only exceptionally,
have produced socio-economic developmental consequences.
Although strengthening of these activities is recommended
since they provide training, information and education, the
intended program will require other strategic efforts, such as
the identification of potential biotechnological applications,
market demands, mechanisms to facilitate the cooperation among
countries, between public and private institutions, to facilitate funding and investments. Institutions with this professional profile do exist and should play a major role in the
program. One should not forget that 75% of the investment in
biotechnology today in the United States is performed by private companies which show a clear tendency to increase these
investments.

TECHNICAL, NATIONAL GLOBAL AND PRIORITIES

One can consider that a technology which results in 20%
increase in productivity leading to a product with a market
share, and which contributes to decreased pesticide consumption will be attractive technically, economically and environ-
mentally. Additionally, these investments must create new job
opportunities and as a consequence contribute to the improvement of the quality of life, particularly in developing countries, as well as in developed countries. One must remember
that 24% of the active population in the United States earns
less than 1/2 of the national per capita income, and that 1
child in 12 in the United States is hungry (in California this
ratio is 1:8). The issue of wealth distribution is not a
"privilege" of developing countries. The criteria for selecting priorities must then consider national priorities. There
are global priorities which are undisputable, such as nitrogen

fixation in grasses, considering the predictable oil shortage
in the future and the lack of its derivatives such as chemical
nitrogen fertilizer. It is possible to establish criteria,
add guidelines to select global and technical priorities for
the program, and to match these to national priorities, as
described below.

The Amazonia in Brazil covers 400 million hectares, close
to 1 billion acres. From that, 10% (e.g. 40 million hectares)
have been destroyed and needs to be recovered. Reforestation
of rain forests which have been destroyed to that extent is a
national as well as a global priority. The genetic resources
as well as the biotechnology to accomplish this task using
African Oil Palm and other palm species is available in Brazil
and a program has been developed on a smaller scale in other
countries such as Indonesia and Malaysia. Technology transfer
from France and/or the United Kingdom, private and public companies for the large scale tissue culture production of propa-
gative material is necessary and can be negotiated as it has
been with other countries. The achievement of such a goal
fulfills two national priorities in Brazil which are the
implementation of an effective land reform project and to
raise the income of 40% of its population (today which earns
less than $50 per month, 1/4 of the national per capita
income) to a level of 200 to 300 dollars per month (based on
the experiences of previously mentioned countries). The project also fulfills two global priorities which are the reforestation of rain forest and the production of an alternative to
diesel oil. Such a project can be accomplished in 40 years at
a cost of $5000 per hectare. The investment needed reaches
$200 billion, almost twice the external debt of Brazil. If
Brazil could attract investments for such a project, the task
of paying the external debt (another global priority) would be
facilitated.

African Oil Palm hybrid clones tested for high oil productivity and propagated by *in vitro* techniques yield from 6 to
12 tons of oil per hectare. At an estimated cost of $500 per
ton, one hectare generates $3000 per year after the third
year. This means that the Brazilian portion of the Amazonian
forest that has been destroyed and is useless today, could

generate $120 billion per year in 40 years (e.g. the external debt of Brazil).

This is just an example to show how technical, national and global priorities can be combined to meet the expectations and the needs of developing and developed countries. Certainly the reforestation of the Amazon will in fact be accomplished using several palm and other native species for environmental safety reasons. The numbers are an exercise that should not differ much if other plants are considered.

Literature Cited

Botstein, D., R. L. White, M. H. Skolnick and R. W. Davis. 1980. Construction of a genetic linkage map in man using restriction fragment length polymorphisms. *Am. J. Hum. Genet.* 32: 314-331.

Burrill, G. S. and K. B. Lee. 1990. *Biotech 91: A changing environment*. Ernst & Young 187. 189 pp.

Feldmann, K. A., M. D. Marks, M. L. Christianson and R. S. Quatrano. 1989. The dwarf mutant of *Arabidopsis* generated by DNA insertion mutagenesis. *Science* 243: 1351-1354.

Mariani, C., M. De Beuckeleer, J. Truetner, J. Leemans and R. B. Goldberg. 1990. Induction of male sterility in plants by a Chimaeric ribonuclease gene. *Nature 347: 737-741*

UNEP. 1990. *Ad hoc Working group of experts on Biotechnology diversity. Relationship between Intellectual Property Rights and Access to Genetic Resources and Biotechnology.*

Williams, J. G. K., A. R. Kubelik, A. J. Livak, J. A. Rafalski and S. C. Tingey. 1990. DNA polymorphisms amplified by arbitrary primers are useful as genetic markers. *Nucleic Acid Res.* 18: 6531-6535.

FEASIBILITY OF OBTAINING COMPARATIVE GENE SEQUENCE DATA FROM PRESERVED AND FOSSIL MATERIALS

David E. Giannasi

Department of Botany, University of Georgia
Athens, GA 30602 U.S.A.

Summary - Until recently most analyses of nucleic acid for
restriction and sequence analyses have required the use of
fresh plant material. In the past few years, it has been
shown that herbarium specimens dried and preserved under generally accepted parameters and without chemical preservation
often possess recoverable DNA. Some samples contain fragments
of genes or organellar genomes that may be amplified using
polymerase chain reaction (PCR) technology. The PCR technology allows for amplification of gene or genomic sequences from
small amounts of DNA preserved in single leaves or small
amounts of leaf tissue (ca. 0.1 g or less). The amplified DNA
may then be subjected to restriction fragment analysis and/or
direct dideoxy base sequencing. These results suggest that
taxa that are not readily accessible due to seasonal or geographic isolation to collectors may still be examined with
molecular techniques using small amounts of herbarium material
or properly stored dried leaf or other tissue. PCR technology
has also recently allowed the recovery and sequencing of organellar DNA from Miocene leaf compression fossils from the
Clarkia Flora of Northern Idaho, U.S.A. Isotopically dated at
ca. 17 MYBP, these Miocene compressions are remarkably preserved in color, shape, dimensions and ultrastructure. This
opens yet another field and option in plant phylogeny allowing
phylogenetic analysis in vertical geological time.

INTRODUCTION

The routine use of chemical constituents preserved in
dried herbarium specimens as taxonomic markers extends back at
least 100 years most notably its use in the systematics of the
lichens (Hale, 1983). The fact that many of these secondary
metabolites, most notably phenolics, alkaloids and other non-

volatile compounds also were preserved in herbarium specimens
of higher plants as well not only indicated their stability
during preservation but provided a potentially large reservoir
of chemical information for taxa which were difficult to
obtain to fill out large surveys of various taxa (Harborne and
Turner, 1984; Giannasi and Crawford, 1986). Indeed, both
older phenolic studies (Bate-Smith, 1962; 1968) and recent
ones (Giannasi, 1978; 1986) owe their comprehensiveness to the
availability of this data in herbarium storage. Sensitive
analytical techniques such as high performance liquid chroma-
tography required even less tissue sample than macro chromatographic techniques and further reduced the need to use large
amounts of material with its concomitant damage to the herbarium specimen.

While herbarium and other dried material were regularly
used for such secondary metabolite surveys (excepting those
compounds quickly lost through volatilization and drying),
little attention was given to the possibility that macromolec-
ular compounds of systematic interest might also be preserved.
Earlier studies of comparative amino acid sequencing by Boul-
ter and colleagues required fresh plant materials (see Harborne and Turner, 1984). The same seems to be true in more
recent times in studies of n-terminal amino acid studies of
specific genes (Martin and Dowd, 1989). On the other hand,
even if it had been shown in the 1960's and 1970's that DNA
was preserved in dried or herbarium material, the technology
was simply lacking to isolate specific gene sequences or organellar genomic sequences for analysis. In the late 1970's and
early 1980's the development of restriction endonucleases that
could digest genomic sequences at specific points provided for
electrophoretic analysis of fragment patterns between species
using Southern blot analyses (Crawford, 1990). The development of convenient dideoxy base sequencing along with plasmid
subcloning procedures allowed for isolation and sequencing of
specific genes by the early 1980's (Ritland and Clegg, 1987;
Palmer et al., 1988). By the late 1980's the practical development of the polymerase chain reaction (PCR) technology had
enabled much of the subcloning procedures to be bypassed in
many types of molecular studies. Using small pieces of

nucleotide sequence of a known gene sequence, called primers,
to "seek" out its related gene sequence within a crude extract
of DNA and allow for the replication and amplification of the
specific gene enabled direct isolation of a gene for sequencing (Innis et al., 1990). It is now possible to use these
primers to specifically locate, replicate and amplify whole
portions or entire organellar genomes which can than be submitted to restriction fragment analysis, bypassing the tedious
Southern blot procedures (Vilgalys and Hester, 1990). This
means that a specific primer can locate a specific gene
sequence from crude DNA (e.g. higher plants) even if it is
contaminated with non-related DNA (bacterial or fungal) since
the latter may lack the target genes (e.g. chloroplast) or do
not have sequences homologous enough to amplify with the primers. Most significant, only the smallest amounts of crude DNA
are needed to survey, given the proper primer and PCR. It is
the latter point which makes the routine use of small amounts
of herbarium material feasible for molecular analyses. Similarly, if DNA preservation occurs in fossil materials, such as
leaf compressions, and is relatively unaltered, it is entirely
possible to perform genetic analysis on any DNA preserved in
fossils as well. This latter possibility adds considerable
potential to molecular systematics and allows phylogenetic
analysis in geological time.

MOLECULAR STUDIES OF DRIED CONTEMPORARY PLANTS

A number of studies exist that have sampled for preservation of secondary metabolites under various regimes (Coradin
and Giannasi, 1980; Cooper-Driver and Balick, 1979; Jacquin-
Dubreuil et al., 1990). Rogers and Bendich (1985) indicated
in their study that randomly chosen samples of herbarium
samples often contained intact and recoverable DNA. To be
sure, not all herbarium samples yielded DNA and many of those
that did, failed to possess fragments of systematically usable
size, the normally expected autodigestion haven taken place
during drying process. However, one might expect some 5-20%
of specimens sampled to possess usable DNA. And, in fact,
this approach has been used regularly and successfully by
Doyle and colleagues for various restriction fragment as well

as sequencing studies (Doyle and Dickson, 1987). Quality of
DNA preservation undoubtedly has much to do with method of
preservation. Doyle and Dickson (1987) and Pyle and Adams
(1989) have also systematically sampled DNA in variously preservation regimes and clearly showed preserved DNA could be
used for restriction analysis. Basically, the DNA was shown
to begin to degrade within a few months. Thus, the ideal time
period to employ dried material is within the first year of
storage. Storage in the cold (e.g. -70°C) was even more advan-
tageous. Indeed, even samples of fresh material experiencing
several weeks in transit before cold storage still yielded
quite adequate material for restriction analysis. These
results suggest that the material may be equally effective
(i.e. dried and prolonged fresh storage) for sequencing work
as well.

The work on the secondary metabolites, however, also shows
that chemical preservation of specimens in the field ultimately alters and/or destroys the secondary metabolites and
the same results were observed on DNA samples in their study
as well. The use of microwave drying (Jacquin-Dubreuil et
al., 1989) is particularly destructive of DNA (Pyle and Adams,
1989). Air or heat dried tissue in the press or in a bag
seems to have a greater chance of containing usable DNA.
Doyle and Dickson (1987) suggest that simple drying halts mold
and fungal destruction more than anything else. From the
author's personal experience, freeze-dried material (frozen in
liquid nitrogen and dried under vacuum for 24 hours) has provided adequate DNA for comparative sequencing studies using a
chloroplast gene such as ribulose-1, 5-bisphosphate carboxy-
lase, large subunit, *rbcL*, for phylogenetic studies in the
angiosperms (Giannasi *et al.*, in press). Indeed, rapid
removal of tissue water by the pressing blotters along with
heat denaturing (but not charring) may also withdraw metabolic
water necessary for autolytic reactions. Circumstantial evidence that this is a probable course of preservation derives
from some of our fossil biochemical work in which fossil leaf
compressions have been preserved in a "natural plant press" of
warm volcanic ash (Niklas and Giannasi, 1977b, see discussion
below).

MOLECULAR ANALYSIS OF FOSSIL LEAF COMPRESSIONS

Considerable work is available discussing the chemical analysis of organic constituents of fossilized plants and compression material both in applied and pure research (Brooks, 1981; Niklas, 1980; 1981; Niklas et al., 1982) and some remarkable chemical phylogenetic work has been accomplished on plant groups at higher taxonomic levels (Niklas and Gensel, 1978). The possibility of finding DNA preserved in fossils of any age has been little considered, the assumption being that such constituents would not survive storage as is suggested by the work on fresh plant material that has been dried. In fact, DNA has been extracted and sequenced from animal and human tissues preserved in bog deposits and dated at 13,000 YBP (Paabo, 1985; 1989; 1990; Paabo et al., 1988; Hagelberg et al., 1989; Higuchi et al., 1984). The same should be true for plants if organic remains are preserved especially as leaf compressions where considerable original tissue is preserved. Indeed, Rogers and Bendich (1985) obtained nanogram amounts of DNA suitable for restriction analysis from plant samples ranging from 22 years to 44,000 years old and work by Helentjaris (1988) suggests its potential in plant analyses from archeological sites.

Such is the case for the Miocene angiosperm leaf compression fossils of the Clarkia Flora of Northern Idaho (U.S.A). Dated by K/A ratios at ca. 17+/-MYBP these leaf compressions are preserved in situ in relatively undisturbed lacustrine shales of an ancient lake formed by volcanic ashfall blockage of the ancient St. Mary's River Drainage (Smiley, 1985; 1989). These shales can be easily pried apart exposing whole leaf macrofossil compressions. Many of these Miocene leaves still possess natural colors including reds, browns and dark greens (Giannasi, 1990). Ultrastructural work on these compressions shows the internal tissue to be well preserved including intact tissue layers, vascular tissue and intracellular organelles including chloroplast membranes (Niklas et al., 1985, W. Thomson, in progress). In addition, a number of internal chemical constituents, including flavonoid pigments, fatty acids, steranes, etc. were also preserved and these were used for chemosystematic purposes to elucidate systematic relation-

ships between fossil taxa and their extant relatives (Giannasi
and Niklas, 1981; 1985; Niklas and Giannasi, 1985; Niklas et
al., 1985). The preservation of leaf and animal remains (fish
and insects) with minimal bacterial decomposition suggested
that the leaves and animals were preserved in a cold, deep,
anoxic lake with little current or upwelling and little bacterial action (Smith and Miller, 1985; Smith and Elder, 1985;
Niklas et al., 1985). Most striking was that different plant
species only a few layers apart still possessed secondary
metabolite patterns distinct from each other and often similar
or identical to an extant relative. This suggested that per-
colation was not occurring in the sediments since the matrix
and all fossils might be expected to have a homogeneous pattern. The presence of these chemical constituents, especially
the flavonoid pigments, was independently documented by Rieseberg and Soltis (1987).

ANALYTICAL PROCEDURES

Unlike the relatively straightforward and controllable
procedures for working with dried or herbarium materials
(Doyle and Dickson, 1987; Pyle and Adams, 1989), isolation and
analysis of fossil materials is largely influenced by the mode
of preservation of the materials. There are a number of
recently studied paleobotanical sites that possess organic
compression material of leaves, flowers and fruits of primarily angiosperms and are likely candidates for such work
(e.g. Crepet and Nixon, 1989; Miller, 1989; Taylor, 1988;
Taylor and Crepet, 1987). Earlier work by Niklas and Giannasi
showed the green leaf compression fossils of the Miocene Succor Creek Flora (Oregon) to have a remarkable degree of ultrastructural preservation (Niklas et al., 1978). The compressions found in this site (Graham, 1963) are preserved in volcanic ash (Fig. 1). The green color is the result of chlo-
rophyll derivatives (Niklas and Giannasi, 1977a,c) which do
not lose their color upon exposure to the atmosphere. The
matrix in which they are found is very hard, requiring a
geologist's hammer to split the layers to expose the plants.
Diagenetic simulations of compression fossilization (Niklas
and Giannasi, 1977b) suggest that the leaves probably exper-

ienced a gradual covering by warm volcanic ash not over 80°C
and at a pH of near neutral. Their physical state shows
remarkable three-dimensionality without crushing and extensive
compression of internal cell structure. This is characteristic of a slow drying process not unlike that of preserved,

Figure 1. Photograph of the original fossil leaf compression
of a *Zelkova* sp. found in a specimen from the Succor Creek
Flora (Graham, 1963), used in the original phytochemical
studies by Niklas and Giannasi (1977). The fossil compression
is approximately 1 inch long and is compared with a leaf of
one of the extant relatives, *Zelkova serrata* (Thunb.) Makino.

desiccated flowers dried in sand or silica gel, which in fact,
aids in retention of original colors as well (see discussion
by Adams *et al*. this volume). Indeed, there appears to be no
bacterial degradation and ultrastructural analysis shows the
preservation of chloroplast membranes with minimal distortion

and autolytic degradation (Fig. 2; Niklas *et al.*, 1978; Niklas *et al.*, 1985). In a strict sense, the preservation of the Succor Creek leaf compressions probably most closely approximates that which occurs when plants are dried in a plant press between newspapers, blotters and under heat, i.e., gentle compression and dehydration and removal of physiological water with subsequent desiccating denaturation of autolytic proteins (Niklas and Giannasi, 1977b). Simulated experiments involving repeated sampling of fresh material over a 48 hour drying

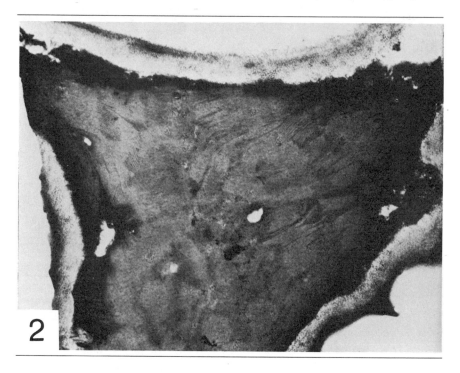

Figure 2. Transmission electron micrograph section through a chloroplast of the fossil *Zelkova* sp. shown in Fig. 1. Note that stacks of thylakoid membranes can be seen in the chloroplast stroma. Magnification ca. 30,000 X. Micrograph courtesy of K. J. Niklas, Cornell University.

period under these conditions (i.e. in a plant press under heat) would provide the precise causes and modes of preservation beyond those outlined in Doyle and Dickson (1987) and as suggested in the fossil work by Niklas and Giannasi (1977b). The excellent preservation of the Succor Creek leaves suggest that some of these fossils may possess recoverable DNA. However, the rock-hard matrix makes collections difficult.

This is not so in the Miocene Clarkia Flora of Northern Idaho some 300 miles further north, near Moscow, Idaho. Although approximately the same age as the Succor Creek Formation, the Clarkia compressions are preserved in soft sediments that are easily pried apart with a knife (Fig. 3). Most

Figure 3. Photograph showing the distinct layers of the lacustrine shale sediments making up part of the lake bed of the Clarkia P-33 main site on the property of Mr. Francis Keenbaum. Large pieces of sediment are pulled out with a pick and then can be gently and easily pried apart with a six inch knife along the planes of the sediment, readily exposing many intact macrofossil compressions.

startling is that many of these leaves also exhibit red, brown
or blackish green colors one would expect from the leaf in its
natural state (Smiley, 1985; 1989; Smiley and Rember, 1985).
Many of the leaves are highly compressed (Fig. 4) but some can
be gently pried off the matrix as a soft fleshy leaf (e.g.
Smilax) while others must be scraped off the matrix (e.g. many
Hamamelidae). In other cases the leaf will dry out and begin

Figure 4. A fossil leaf compression of a *Zelkova* sp. from
Clarkia. Compare with the *Zelkova* species from Succor Creek
shown in Fig. 1. Actual fossil size is ca. 3 in long.

to lift off exposing original leaf material (Fig. 5). Many
such taxa show remarkable ultrastructural preservation (Niklas
et al., 1985; W. Thomson, in progress). Equally startling is
that many of the leaves exhibit a layer of water on the surface (Giannasi, 1990). Examination of the sediment matrix
fails to show contamination with natural products from the

leaves and as indicated, leaves only a layer or two away show

Figure 5. A fossil leaf compression of *Magnolia latahensis* (Berry) Brown from Clarkia. The original color was a dark reddish brown fading to black. Note that the leaf is lifting from the shale matrix. Actual size ca. five inches.

a taxonomic chemical profile distinct from the nearest neighbor. Both of these facts mediate against percolative contamination, suggesting that the surface water may indeed represent the original water accompanying burial and compression of the leaf.

Unlike the Succor Creek compressions, the Clarkia leaves do not retain their color permanently after exposure. Within minutes of exposure, the compressions turn black. However, previous work indicates that the secondary metabolites do not decompose(Giannasi and Niklas, 1981; Niklas *et al.*, 1985) nor is the ultrastructure destroyed (Fig. 6). This was first thought to be the result of rapid oxidation due to exposure to the atmosphere. However, crude experiments exposing specimens in a reducing atmosphere under nitrogen or inert gases fails

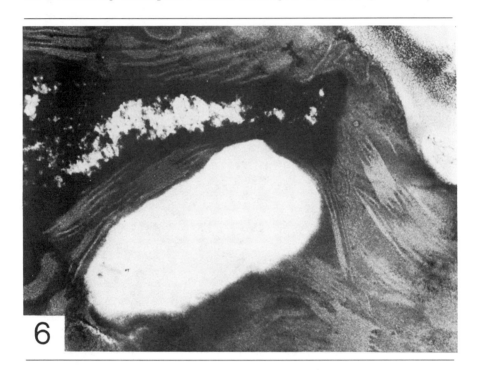

Figure 6. Transmission electron micrograph through the chloroplast stroma of a leaf compression of a fossil *Betula* sp. from the Clarkia Flora. Note well preserved thylakoid stacks in the lower right hand corner of the micrograph. Approximate magnification is 38,400 X. Micrograph courtesy of K. J. Niklas, Cornell University.

to halt the reaction. It appears, therefore, that the reaction, if it is oxidative, has the capability possibly stored up in the attendant water surface or cellular water (or other substances) "created" during the preservation process. What is truly amazing is that secondary metabolites and even DNA is preserved in the tissue despite these overt reactions and have not been leached out or undergone autolytic action. Certainly this type of preservation warrants attempts at simulated diagenesis under similar artificial conditions.

The possibility that DNA was also preserved in these leaf compressions was also discussed by the author with others but suitable technology for (i.e., PCR) for working with the small amounts of fossil tissue were not available until the late 1980's. The advent of practical PCR technology in the late 1980's allowed the author and his colleagues to return to the Clarkia site to collect some compression samples. The techniques of Rogers and Bendich (1985) were used on the fossil material and small amounts of DNA were extracted from at least 40% of the samples (Golenberg et al., 1990). The large subunit of the RuBisCo gene (rbcL) of the chloroplast had already been sequenced in comparative studies of extant plants (Giannasi et al., in press). PCR primers used in these studies were used to probe the fossil DNA for fragments of this gene. Indeed, one compression specimen of Magnolia latahensis (Berry) Brown, yielded by PCR amplification, a 820 base fragment of rbcL which was sequenced and compared with sequences from extant Magnolia macrophylla L. and Liriodendron tulipifera L., both of the Magnoliaceae, as well as sequences from a number of different angiosperm subclasses (see Fig. 3 in Golenberg et al., 1990). The partial sequence of fossil Magnolia clusters readily with the other taxa in the Magnoliaceae. Some variation in clustering is observed within the Magnoliaceae in that the fossil Magnolia latahensis clustered with Liriodendron in some of the analyses. This is probably due to the fact that only 770 of the possible 1470 bases for rbcL were available from the fossil for comparison with the full sequence from the extant taxa. Even so, the pairing within the Magnoliaceae is remarkably consistent. An additional 400-500 bases from the fossil would probably add more specificity to the clustering of the fossil Magnolia with its living relative. The same data were submitted to an alternative parsimony analysis (Fig. 7) using PAUP (Swofford, 1990). The Magnolia cluster obtained using parsimony is identical to that obtained by Golenberg et al. (1990) using a phenetic system (UPGMA).

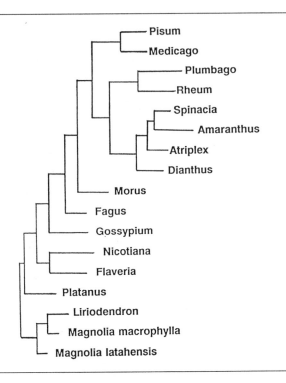

Figure 7. A cladistic analysis of *rbcL* sequences of a fossil
compression of *Magnolia latahensis* (Berry) Brown with a
related living species of *Magnolia* and other representative
taxa from the Magnoliaceae (e.g. *Liriodendron*) and a number of
outgroup dicot taxa. PAUP 3.0 (Swofford, 1990) was employed;
the tree represents a consensus of several similar trees all
of ca. 814 steps. Note that the clustering of the fossil *Magnolia* in the Magnoliaceae clade is identical to that obtained
by Golenberg *et al.* (1990, Fig. 3) using a phenetic analysis
(UPGMA). The out group taxa also fall into taxonomically
recognized clades similar or identical to those observed in
other studies of extant taxa (Giannasi *et al.*, in press;
Palmer *et al.*, 1988). The full scientific name for each of
the species sampled in this parsimony analysis is given in
Golenberg *et al.* (1990) and Giannasi *et al.* (in press).

These results indicate that the original identification based on morphology was correct and that such sequencing comparisons may be effectively used to confirm generic identifications of fossil taxa. In a genus with many species, relating a fossil taxon to its nearest living relative, even if all 1470 bases are available, becomes more difficult as the number of living species increases. The highly conserved nature of *rbcL* would probably not provide enough nucleotide changes for distinguishing between 30 or more extant species. *rbcL* sequences of three tobacco species show less than 10-15 bases different between three species. Thus, using *rbcL* a fossil species can be identified to the generic level. At the interspecific level, a gene or part of a genome with considerably more variation would be more effective in separating large numbers of species once their generic affinities have been established. There is some evidence to suggest that portions of non-coding regions prior to or after the end of *rbcL* provide such increased ranges of nucleotide variation that would aid interspecific differentiation (Golenberg, in press).

DOCUMENTATION OF FOSSIL DNA STUDIES

One of the difficulties in working with the fossils is that unlike extant plants there is little possibility of retaining compression material as a voucher as would be the case in purely morphological studies. To date, basic extraction of the fossil material has taken place at the site to avoid as much as possible any accelerated decomposition of the DNA due to any unknown degradative processes (Golenberg et. al. 1990; Giannasi, 1990). Upon exposure of the fossil and primary identification by expert paleobotanists at the site, the fossil is immediately photographed on slide film. This is really the only documentation of the macrofossil unless a morphologist is present to preserve a portion of the cuticle for corroboration. In many cases, the whole leaf must be used to obtain adequate DNA. In other cases the identity of the fossils (e.g. *Magnolia*, *Persea*, *Liriodendron*, *Quercus* sp., *Fagus*, etc.) and detailed identification has already been carried out (Smiley and Rember, 1985) or their identification by gross morphology is relatively obvious (Fig. 4).

Once the DNA has been extracted, it can be probed (Southern blot analysis) or subjected to direct PCR amplification with primers specific to individual genes (Golenberg et al., 1990) or genomic sequences (Vilgalys and Hester, 1990), or even to retroviral elements (M. Cummings, Harvard University). Indeed, considering progress with analysis of various genes, both organellar and nuclear, PCR provides the most sensitive test to efficiently locate known, phylogenetically useful target DNA sequences. Like all other plant materials however, the amount of DNA extracted is limited as are the fossil specimens. New specimens can be reisolated and extracted but this also implies that the extent of the fossil site must also be determined (e.g. using sample coring) if the site is not to be "used up." Individual scientists may visit the site to collect but more efficient would be the establishment of a registry of available DNA for various taxa. This would be of particular significance since several different independent DNA sequences could eventually be compared for phylogenetic and evolutionary studies. The University of Idaho, for example, has established the Tertiary Fossil Research Center, Moscow, Idaho, currently administered by Dr. C. J. Smiley, to monitor and develop the Clarkia Site now and in the future.

CORROBORATIVE STUDIES BY OTHER LABORATORIES

Several other laboratories have been invited or are currently working independently on the Clarkia fossil compressions as part of a program to corroborate the original studies by Golenberg et al. (1990). Drs. Douglas and Pamela Soltis, Washington State University, have successfully amplified a fragment of rbcL from fossil leaf compressions of Taxodium from Clarkia and the sequence clusters well with a putative modern relative, Metasequioia. Also, Dr. Holly A. Wichman and colleagues, University of Idaho, have amplified a fragment of rbcL from a fossil compression of Quercus from Clarkia. Laboratories at several universities have successfully and independently isolated retrotransposons in several fossil samples and their related extant relatives (Golenberg, in press). One laboratory (Paabo and Wilson, 1991) has failed to duplicate the results of Golenberg et al. (1990) and has expressed some

doubt and criticism over the original work.   These criticisms
and doubts (e.g. Paabo *et al.*, 1990) have been addressed
experimentally in detail by Golenberg (in press).   Considering
recent repeatable successes in other independent laboratories
the commentary by Paabo and Wilson (1991) may be somewhat premature.   Indeed, the results of DNA analyses of the Clarkia
materials hopefully will encourage paleobotanists and geolo-
gists to reexamine their sites, both of younger and older
ages, for the possible existence of other compression samples
which may contain preserved nucleic acids.

RECOMMENDATIONS FOR USE OF HERBARIUM/FOSSIL DNA

For the use of herbarium material the following may represent reasonable guidelines for preservation and storage of
such material:

1.   Establishment of a central listing agency (e.g. DNA
Bank-Net) which will have a register of available laboratories
or institutions which have DNA extracts for one or more
species that may be of interest to investigators.

2.   Registered institutions agree to be responsible for
storage and accessibility of these DNA samples and their
replacement as extracts degrade over time.

3.   Registered institutions also store and maintain at
least one voucher herbarium specimen documenting each plant
extract, the specimen to be stored at the registered institution or at facilities of the central listing agency.

4.   The central listing agency provide a written protocol
upon request to institutions interested in participation in
the program that provide a uniform procedure for extraction
and storage of the data.

5. Copies of all research resulting from use of all listed
extracts should be provided to the central listing agency and
the donating institution, which should receive acknowledgment
for its use.

For the use of fossil material many of the above recommendations are applicable including items 1, 2, 4 and 5.   Provision of a voucher specimen (item 3) may represent more of a
problem since in most cases most of the fossil specimen may
used in the analysis.   However, an accurate photograph, pre-

ferably in black and white in addition to any color slides or
prints would be sufficient record as a voucher. Indeed, since
many institutions no longer exchange many of their more rare
specimens the use of a photograph or a microfiche has become
an acceptable mode of documentation. If possible, a small
piece of the leaf or flower preserved in solution or on a
slide for future analysis (e.g. leaf venation) and as a partial voucher is also to be recommended. Perhaps most significant in the use of fossil materials is the need of accurate
dating of the material and provision of adequate documentation
of the preservation regimes resulting in the particular fossil
specimen.

All too often, many plant species are rarely collected or
difficult to collect if in relatively inaccessible areas. The
possibility of using herbarium specimens to provide DNA for
molecular analyses seems most reasonable. At the same time it
must be remembered that herbarium specimens also represent
limited resources and thus all current policies applied by
herbaria for use of such materials must also apply for sampling for molecular work. Certainly, the ability to use dried
material in these studies provides a considerable flexibility
to the use of field collected material. Within the time limits outlined by Doyle and Dickson (1987), Pyle and Adams
(1989) and Adams et al. (this volume), such an approach also
provides easier collection and storage of such dried material
by field taxonomists without any need to involve themselves in
any actual extraction work.

With decreased funding and travel funds available for collecting, plant diversity offensives not withstanding, and the
increasingly rapid destruction of plant species as countries
experience their manifest destiny of development, the herbarium and its potentially enormous store of data for molecular
studies obviously emphasizes the long-standing value of these
collections. This does not mean uncontrolled access to herbaria, but these collections do have a contribution to make as
part of the DNA data bank outlined in the provisions discussed
above. Equally as significant, the same applies to paleobotanical collections as well.

Acknowledgments - My thanks to Dr. Edward M. Golenberg, Wayne
State University for allowing me to use some of his most
recent sequence data results from his most recent studies of
the Clarkia Fossils, to Dr. Gerald Learn, University of Cali-
fornia-Riverside for assistance with the cladistic analyses
and Dr. Michael Clegg, University of California, Riverside,
whose appreciation of technology and the potential of the
Clarkia site long ago brought much of this work to fruition.
My thanks also to my friend and colleague, Dr. Karl Niklas,
Cornell University, for access to several of the illustrations
of fossil ultrastructure. And as always my appreciation to
Mr. Francis Keenbaum, for generous access to the fossil site
to continue my studies on the Clarkia fossils, and to Dr. Jack
Smiley, Acting Director, Tertiary Fossil Research Center, University of Idaho, Moscow, ID, and his wife, Peg Smiley, for
their friendship and unflagging hospitality and aid on the
fossil digs at Clarkia. This work has been supported by funding from the Department of Botany, University of Georgia, the
Office of the Vice President for Research, University of Georgia and from a M. G. Michael Research Award (1990), University
of Georgia, awarded to the author.

Literature Cited

Bate-Smith, E. C. 1962. The phenolic constituents of plants
     and their taxonomic significance. I. Dicotyledons. *J.
     Linn. Soc. Bot.* 58: 73-95.
_____. 1968. The phenolic constituents of plants and their
     taxonomic significance. II. Monocotyledons. *J. Linn. Soc.
     Bot.* 60: 325-356.
Brooks, J. (ed.). 1981. *Organic maturation studies and fos-
     sil fuel exploration.* Academic Press, London.
Cooper-Driver, G. A. and M. A. Balick. 1979. Effects of field
     preservation on the flavonoid content of *Jessenia batava.*
     Bot. Mus. Leafl. 26: 257-265.
Coradin, L. and D. E. Giannasi. 1980. The effects of field
     preservation on plant collections to be used in chemotaxo-

nomic surveys. *Taxon* 29: 33-40.

Crawford, D. J. 1990. *Plant molecular systematics*. Wiley Interscience, New York.

Crepet, W. L. and K. C. Nixon. 1989. Extinct transitional Fagaceae from the Oligocene and their phylogenetic implications. *Amer. J. Bot.* 76: 1493-1505.

Doyle, J. J. and E. E. Dickson. 1987. Preservation of plant samples for DNA restriction endonuclease analysis. *Taxon* 36: 715-722.

Giannasi, D. E. 1978. Generic relationships in the Ulmaceae based on flavonoid chemistry. *Taxon* 27: 331-334.

_____. 1986. Phytochemical aspects of phylogeny in Hamamelidae. *Ann. Missouri Bot. Garden* 73: 417-437.

_____. 1990. The Clarkia "fossil" leaves: windows to the past. *Northwest Science* 64: 232-235.

_____ and D. J. Crawford. 1986. Biochemical systematics. II. A Reprise. Pp. 25-248. In: M. K. Hecht, B. Wallace and G. T. Prance (eds.), *Evolutionary biology*. Plenum, New York.

_____ and K. J. Niklas. 1981. Comparative paleobiochemistry of some fossil and extant Fagaceae. *Amer. J. Bot.* 68: 762-770.

_____ and _____. 1985. The paleobiochemistry of fossil angiosperm floras. Part I. Chemosystematic aspects. Pp. 161-174. In: C. J. Smiley (ed.), *Late Cenozoic history of the Pacific Northwest*. AAAS (Pacific Division), San Francisco.

_____, G. Zurawski, G. Learn and M. T. Clegg. 1991. Evolutionary relationships of the Caryophyllidae based on comparative *rbcL* sequences. *Syst. Botany.* (in press).

Golenberg, E. M. 1991. Amplification and analysis of Miocene plant fossil DNA. *Proc. Soc. Roy.*, London.

_____, D. E. Giannasi, M. T. Clegg, C. J. Smiley, M. Durbin, D. Henderson and G. Zurawski. 1990. Chloroplast DNA sequence from a Miocene *Magnolia* species. *Nature* 344: 656-658.

Graham, A. 1963. Systematic revision of the Succor Creek and Trout Creek Miocene floras of southeastern Oregon. *Amer. J. Bot.* 50: 921-936.

Hagelberg, E., B. Sykes and R. Hedges. 1989. Ancient bone DNA amplified. *Nature* 432: 485.

Hale, M. E., Jr. 1983. *The biology of lichens.* 3rd Ed. Edward Arnold, Baltimore.

Harborne, J. B. and B. L. Turner. 1984. *Plant chemosystematics.* Academic Press, Orlando, Florida.

Helentjaris, T. 1988. Does RFLP analysis of ancient Anasazi samples suggest that they utilized hybrid maize: *Maize Genetics Cooperative Newsletter* 62: 104-105.

Higuchi, R., B. Bowman, M. Friedberger, O.A. Ryder and A. C. Wilson. 1984. DNA sequences from the quagga, an extinct member of the horse family. *Nature* 312: 282-284.

Innis, M. A., D. H. Gelfand, J. J. Sninsky and T. J. White, (eds.). 1990. *PCR protocols, a guide to methods and application.* Academic Press, San Diego.

Jacquin-Dubreuil, A., C. Breda, M. Lescot-Layer and L. Allorge-Boiteau. 1990. Comparison of the effects of a microwave drying method to currently used methods on the retention of morphological and chemical leaf characters in *Nicotianum tabacum* L. cv. *Samsun. Taxon* 38: 591-596.

Martin, P. G. and J. M. Dowd. 1989. Phylogeny among the flowering plants as derived from amino acid sequence data. Pp. 195-204. In: B. Fernholm, K. Bremer and H. Jornvall (eds.), *The hierarchy of life.* Elsevier, Amsterdam.

Miller, N. G. 1989. Structurally preserved leaves of *Harrimanella hypnoides* (Ericaceae): paleoecology of a new North American Pleistocene fossil. *Amer. J. Bot.* 76: 1089-1095.

Niklas, K. J. 1980. Paleochemical techniques and their applications to paleobotany. Pp. 143-181. In: L. Reinhold, J. B. Harborne and T. Swain (eds.), *Progress in phytochemistry*, Vol. 7. Pergamon, Oxford.

_____. 1981. Implications on pre- and post-depositional environments of ultrastructural-biochemical correlations in fossil angiosperm leaf tissues. Pp. 111-144. In: J. Brooks (ed.), *Organic maturation studies and fossil fuel exploration.* Academic Press, London.

_____ and P. G. Gensel. 1978. Chemotaxonomy of some Paleozoic vascular plants. Part III. Cluster configurations and their bearing on taxonomic relationships. *Brittonia* 30: 216-232.

_____ and D. E. Giannasi. 1977a. Flavonoids and other chemical constituents of fossil Miocene *Zelkova* (Ulmaceae). *Science* 196: 877–878.

_____ and _____. 1977b. Geochemistry and thermolysis of flavonoids. *Science* 197: 767–769.

_____ and _____. 1977c. The green fossils. *Garden* 1: 6–15.

_____ and _____. 1985. The paleobiochemistry of fossil angiosperm floras. Part II. Diagenesis of organic compounds with particular reference to steroids. Pp. 175–183. In: C. J. Smiley (ed.). *Late Cenozoic history of the Pacific Northwest*. AAAS (Pacific Division), San Francisco.

_____, _____ and N. L. Baghai. 1985. Paleobiochemistry of a North American fossil *Liriodendron* sp. *Biochem. Syst. Ecol.* 13: 1–4.

_____, R. M. Brown, Jr. and R. Santos. 1985. Ultrastructural states of preservation in Clarkia angiosperm leaf tissues: implications on modes of fossilization. Pp. 143–159. In: C. J. Smiley (ed.). *Late Cenozoic history of the Pacific Northwest*. AAAS (Pacific Division), San Francisco.

_____, _____, _____ and B. Vian. 1978. Ultrastructure and cytochemistry of Miocene angiosperm leaf tissues. *Proc. Natl. Acad., USA* 75: 3263–3267.

_____, B. H. Tiffney and A. C. Leopold. 1982. Preservation of polyunsaturated fatty acids in Paleogene angiospem fruits and seeds. *Nature* 296:63–64.

Paabo, S. 1985. Molecular cloning of ancient Egyptian mummy DNA. *Nature* 314: 644–645.

_____. 1989. Ancient DNA: extraction, characterization, molecular cloning, and enzymatic amplification. *Proc. Natl. Acad., USA* 86: 1939–1943.

_____. 1990. Amplifying ancient DNA. Pp. 159–166. In: M. A. Innis, D. H. Gelfand, J. J. Sninsky and T. J. White (eds.), *PCR protocols: a guide to methods and applications*. Academic Press, San Diego.

_____ and A. C. Wilson. 1991. Miocene DNA sequences-A dream come true? *Current Biology*: 45–46.

_____, J. A. Gifford and A. C. Wilson. 1988. Mitochondrial DNA sequences from a 7000-year old brain. *Nucleic Acids Res.* 16: 9775–9787.

_____, D. M. Irwin and A. C. Wilson. 1990. DNA damage promotes jumping between templates during enzymatic amplification. *Jour. Biol. Chem.* 265: 4718-4721.

Palmer, J. D., R. K. Jansen, H. J. Michaels, M. W. Chase and J. R. Manhart. 1988. Chloroplast DNA variation and plant phylogeny. *Ann. Missouri Bot. Gard.* 75: 1180-1206.

Pyle, M. M. and R. P. Adams. 1989. *In situ* preservation of DNA in plant specimens. *Taxon* 38: 576-581.

Rieseberg, L. H. and D. E. Soltis. 1987. Flavonoids of fossil Miocene *Platanus* and its extant relatives. *Biochem. Syst. Ecol.* 15: 109-112.

Ritland, K. and M. T. Clegg. 1987. Evolutionary analysis of plant DNA sequences. *Amer. Naturalist* 130: S74-S100.

Rogers, S. O. and A. J. Bendich. 1985. Extraction of DNA from milligram amounts of fresh, herbarium and mummified plant tissues. *Plant Mol. Biology* 5: 69-76.

Smiley, C. J. (ed.). 1985. *Late Cenozoic history of the Pacific Northwest.* AAAS (Pacific Division), San Francisco.

_____. 1989. The Miocene Clarkia fossil area of northern Idaho. Pp. 35-48. In: V. E. Chamberlain, R. M. Breckenridge and B. Bonnichsen (eds.), *Guidebook to the geology of Northern and Western Idaho and surrounding area. Idaho Geol. Survey Bull.* 28.

_____ and W. C. Rember. 1985. Composition of the Miocene Clarkia flora. Pp. 95-112. In: C. J. Smiley (ed.), *Late Cenozoic history of the Pacific Northwest.* AAAS (Pacific Division), San Francisco.

Smith, G. R. and R. R. Miller. 1985. Taxonomy of fishes from Miocene Clarkia lake beds, Idaho. Pp. 75-83. In: C. J. Smiley (ed.), *Late Cenozoic history of the Pacific Northwest.* AAAS (Pacific Division), San Francisco.

_____ and R. L. Elder. 1985. Environmental interpretation of burial and preservation of Clarkia fishes. Pp. 85-93. In: C. J. Smiley (ed.), *Late Cenozoic history of the Pacific Northwest.* AAAS (Pacific Division), San Francisco.

Swofford, D. L. 1990. *PAUP: Phylogenetic analysis using parsimony, Version 3.0.* Illinois Natural History Survey, Champaign.

Taylor, D. W. 1988. Eocene floral evidence of Lauraceae: corroboration of the North American megafossil record. *Amer. J. Bot.* 75: 948-957.

_____ and W. L. Crepet. 1987. Fossil floral evidence of Malphigiaceae and an early plant-pollinator relationship. *Amer. J. Bot.* 74: 274-286.

Vilgalys, R. and M. Hester. 1990. Rapid genetic identification and mapping of enzymatically amplified ribosomal DNA from several *Cryptococcus* species. *Jour. Bacteriology* 172: 4238-4246.

PROSPECTS OF ACCESSING DNA BANKS FOR THE ISOLATION OF GENES ENCODING BIOLOGICALLY ACTIVE PROTEINS

Richard A. Dixon and Nancy L. Paiva

Plant Biology Division, The Samuel Roberts Noble Foundation
P. O. Box 2180, Ardmore, Oklahoma 73402 USA.

Summary - The feasibility of utilizing genes from DNA banks is examined. A summary of the efforts to clone genes of phenylpropanoid metabolism is presented. These genes have been cloned from five major phenylpropanoid pathways: central phenylpropanoid, lignin branch, flavonoid branch, isoflavonoid branch and the stilbene branch pathway. Genes have been cloned from the following alkaloid classes: monoterpene indole, indole and isoquinoline, isoquinoline, and tropane. Two other gene classes are reviewed: genes that encode bioactive proteins and regulatory proteins. The major limitation for utilization of DNA from banks does not appear to be technology, but the lack of detailed basic knowledge concerning the biochemistry, bioactivity, and pharmacology of plant chemicals.

INTRODUCTION

A large proportion of the pharmacologically active compounds in use today are of plant origin, and current trends in crop protection strategies indicate a moving away from synthetic agrichemicals to natural products. It is therefore a major concern of plant conservationists that the loss of endangered species in the wild will deprive us of potential cures or treatments for human, animal and plant diseases. Clearly, the predicted rate of species loss is such that an exhaustive screening for biologically active agents, coupled with collection and propagation of their sources, is not feasible. Our present understanding of chemotaxonomy may suggest what type of species should be targeted for such study, but a vast number of species still remain to be classified. Collection of samples for storage in DNA banks may indeed preserve the genetic material of species soon to become extinct, but

the question then arises of how to access genomic DNA for
valuable gene sequences in the absence of knowledge of the
biochemistry and pharmacology of the source.

In the last few years, genes encoding a number of proteins
which either exhibit biological activity themselves, or are
enzymes involved in the synthesis of biologically active secondary products, have been cloned from plants. Thus, DNA
sequence information is available which may aid the design of
probes for isolation of genes with homologous sequences from
previously un-investigated species. In this chapter we present a brief overview of the strategies behind the cloning of
such genes, and discuss how far our current knowledge of plant
molecular biology, and newly developed techniques for gene
isolation, can be applied towards accessing DNA banks for
potentially useful genes.

In Table 1, we present a simple functional classification
of plant genes. Basic biochemical considerations dictate the
potential applied value of newly discovered genes within these
several classes in terms of the types of variants which might
be desirable. Biologically active secondary metabolites
clearly represent one of the major resources of the plant
kingdom. From the conservation view point, the worrying truth
is that our understanding of the enzymatic reactions of plant
secondary product biosynthesis is, with some of the exceptions
reviewed below, still rudimentary. In contrast, variation of
catalytic or biological activity of known enzymes is now
attainable by *in vitro* site-directed mutagenesis. Such
studies may ultimately provide a predictive knowledge base for
the design of novel enzymes. This approach has already been
successful with respect to altering the substrate specificity
of a cytochrome P450 enzyme (Lindberg and Negishi, 1989). In
relation to secondary product metabolism, understanding the
principles of molecular catalysis underlying, for example, the
stereospecific carbon skeleton rearrangements of isoprenoid
compounds, or regio-specific substitution reactions, could
ultimately enable new secondary product pathways to be engineered in recombinant organisms. There is, however, a pressing need to expand the data base by further biochemical investigation of novel secondary product pathways.

Table 1. A functional classification of plant gene products.

Enzymes of primary metabolism:
Function: Potentially desirable variants:

Catabolism, biosynthesis   1. Altered catalytic capacity
of proteins, nucleic acids,   2. Altered allosteric regulation
carbohydrates and lipids.   3. Increased stability
                              4. Increased temperature optimum

Enzymes of secondary metabolism:
Function: Potentially desirable variants:

Biosynthesis of phenyl-   1. Altered catalytic capacity
propanoids, alkaloids,   2. Altered substrate specificity
terpenoids, etc.   (stereo-and regio-specificity)
                              3. Altered transcriptional and
                                 post-translational regulation

Biologically active proteins:
Function: Potentially desirable variants:

Defense   1. Increased activity
                              2. Altered antimicrobial spectrum
                              3. Altered pharmacological
                                 activity

Regulatory proteins:
Function:

Signal transduction, gene regulation.

Structural proteins:
Function:

Cell wall, cytoskeleton.

METHODS FOR GENE ISOLATION

Many of the plant genomic or cDNA sequences obtained to
date have been isolated by the standard practice of screening
a genomic/cDNA library, usually in a lambda phage or plasmid
vector, with a suitable DNA probe or, in the case of cDNA
expression libraries, with an antibody. In view of the relative instability of RNA compared to DNA, collection, transport
and long term preservation of material containing undegraded
transcripts from which cDNA libraries could be constructed
would be much more difficult than preservation of genomic DNA.
Certain screening methods which have been useful for cDNA isolation e.g. antibody screening (Young and Davis, 1983), differential hybridization (Somssich et al., 1989) or production
of cDNA subtraction libraries (Timblin et al., 1990) are pre-
cluded for genomic DNA banks.

Standard methods for genomic library production in phage
lambda vectors (Kaiser and Murray, 1985) are generally applicable to plant DNA. As libraries can be amplified, they present a renewable source of plant species-specific gene
sequences, although some clones may be selectively lost during
repeated amplification, and library titers decrease as a function of time of storage. Whether it is decided to make a
library (which may require up to 100 μg of pure, high molecular weight plant DNA) or to use uncloned genomic DNA as primary source of new genes (see below), suitable DNA probes will
be necessary for screening. These could be either (1) a partial or full-length cDNA, or complementary riboprobe, of a
previously isolated and characterized gene or (2) a set of
degenerate oligonucleotides based on sequence motifs conserved
between known genes encoding the same gene product from other
sources or common to functional regions within a particular
family of genes (e.g. heme binding sites of cytochromes or
ATP-binding sites of protein kinases). The use of two sets of
degenerate oligonucleotides, based on two separate conserved
regions, is often necessary to avoid an unacceptable number of
false positives when screening libraries directly with oligonucleotide probes (Lawton et al., 1989). The extent of degen-
eracy of oligonucleotide probes could be lessened if any
unusual codon usage preferences were known for the new species

under investigation. These could be investigated by first
isolating and sequencing common genes for which sequence
information, and therefore probes, was available in other
species. Such studies would additionally provide valuable
taxonomic and evolutionary information. Occasionally, however, unusual codon usage preferences are observed for specific proteins within one species. For example, the bean
hydroxyproline rich glycoprotein (HRGP) Hyp 2.11 has a different codon usage for proline than the other HPGPs from this
source (Sauer et al., 1990). Plant codon usage has recently
been reviewed (Campbell and Gowri, 1990).

Polymerase chain reaction (PCR) amplification of DNA is a
powerful and extremely sensitive method for the isolation of
specific DNA sequences (Saiki et al., 1988). PCR methods can
be applied to cDNA or genomic libraries either to bypass some
of the time-consuming filter hybridization steps during initial library screening (Bloem and Yu, 1990), or to amplify
directly a specific gene sequence (Edington et al., 1991). A
new application, inverse PCR, allows the geometric amplification of unknown flanking sequences surrounding a core region
of known sequence (Ochman et al., 1990). The sensitivity of
the PCR technique makes it ideally suited to the direct amplification of sequences from small amounts of genomic DNA. Naturally, however, the requirement for oligonucleotide primers
based on prior sequence information is the same as for filter
hybridization of genomic libraries.

Some cloning strategies have been devised which rely on
function of the cloned sequence rather than prior sequence
information. For example, an *Agrobacterium* T-DNA-based vector
containing abbreviated TATA regions from the cauliflower
mosaic virus 35S promoter linked to the β-glucuronidase (GUS)
reporter gene has been used for construction of an *Arabidposis
thaliana* genomic library; transformation of tobacco plants
with individual clones and screening for up-regulated GUS
expression resulted in the identification of five *Arabidopsis*
genomic sequences with enhancer function (Ott and Chua, 1990).
In principle, such transformation-based functional screens
could be used for isolation of novel genes yielding predictable phenotypes, but the approach is extremely labor inten-

sive. Another new technique, whole genome PCR, has been utilized for the cloning of human sequences which bind to *Xenopus*
transcription factor (Kinzler and Vogelstein, 1989). This
strategy can be adapted to isolate sequences that can be
selected by any physical or biological method.

GENES ENCODING ENZYMES INVOLVED IN THE SYNTHESIS OF BIOLOGI-
CALLY ACTIVE SECONDARY METABOLITES

It can be seen from the above considerations that
increased knowledge of the molecular (i.e. protein sequence)
basis of enzymatic catalysis will be necessary if DNA banks
are to be successfully accessed for secondary product biosynthetic genes other than homologs of those already isolated.
It is only in the last few years that any of the genes encoding enzymes of secondary product synthesis have been cloned.
The sections below review selected areas where progress is
being made.

Enzymes of phenylpropanoid biosynthesis - The enzyme L-phenyl-
alanine ammonia-lyase (PAL) diverts the amino acid phenylalanine into a range of phenylpropanoid secondary products
including lignin, lignans, flavonoids, isoflavonoids and stil-
benes. Much progress has been made in recent years on the
isolation of genes encoding enzymes of the central phenylpropanoid pathway and its specific branch pathways (Table 2).
Much of this work has been reviewed (Hahlbrock and Scheel,
1989; Dixon and Harrison, 1990).

The initial cloning of bean PAL (Edwards et al., 1985)
involved differential cDNA hybridization, with confirmation of
identity by hybrid release translation and specific immunopre-
cipitation. The bean PAL cDNA has been shown to cross hybrid-
ize with PAL from a wide range of plant species (including
tobacco, parsley and strawberry), and sequence comparison of
PALs from bean, rice, sweet potato, parsley and alfalfa indicate a number of extensive, highly conserved regions of amino
acid sequence homology. It is therefore likely that PAL genes
could be isolated from new species by heterologous hybridization. Similar extensive sequence homology is seen among the
genes encoding chalcone synthase from a variety of sources

Table 2.   Cloned genes of phenylpropanoid metabolism (as of January 1991).   See Dixon and Harrison (1990) for a detailed review up to mid-1990.

Central phenylpropanoid pathway

Gene	Source
L-phenylalanine ammonia-lyase	*Arabidposis thaliana*
	Daucus carota
	Ipomea batatas
	Medicago sativa
	Oryza sativa
	Phaseolus vulgaris
	Petroselinum crispum
	Solanum tuberosum
4-coumarate: CoA ligase	*Oryza sativa*
	Petroselinum crispum
	Solanum tuberosum

Lignin branch pathway

Gene	Source
Caffeic acid 3-0-methyl-transferase	*Medicago sativa*
	Populus tremuloides
Coniferyl alcohol dehydrogenase	*Phaseolus vulgaris*
Lignin peroxidase	*Nicotiana tabacum*

Flavonoid branch pathway

Gene	Source
Chalcone synthase	*Arabidopsis thaliana*
	Glycine max
	Hordeum vulgare
	Lycopersicon esculentum

(Table 2 continued)

	Magnolia liliiflora
	Matthiola incana
	Medicago sativa
	Petroselinum crispum
	Petunia hybrida
	Phaseolus vulgaris
	Pisum sativum
	Ranunculus acer
	Zea mays
Chalcone isomerase	*Petunia hybrida*
	Phaseolus vulgaris
Dihydroflavanol 4-reductase	*Antirrhinum majus*
	Petunia hybrida
	Zea mays

Isoflavonoid branch pathway

Gene	Source
Isoflavone reductase	*Medicago sativa*

Stilbene branch pathway

Gene	Source
Stilbene synthase	*Arachis hypogaea*

(Niesbach-Klosgen *et al.*, 1987).

In addition to providing heterologous probes for isolation of homologous sequences from other plant sources, cloning of genes encoding enzymes catalyzing other reactions in the phenylpropanoid/isoflavonoid pathway may increase our basic knowledge of the relation between protein structure and enzymatic activity, such that predictive approaches could be taken for the design of probes to recognize sequences encoding novel enzymes from previously uninvestigated sources. For example, alfalfa cells appear to contain a number of 0-methyltransferases of similar biochemical properties but with different sub-

strate specificities. Specific enzymes catalyze the 3-0-
methylation of caffeic acid (lignin pathway), the 2'-0-
methylation of 2',4,4'-trihydroxy-chalcone and the 7 or
4'-0-methylation of isoflavones (Edwards and Dixon, 1991a, b;
C. Maxwell, R. Edwards and R. A. Dixon, unpublished results).
Similarly, different cytochrome P450s catalyze the 4-hydroxy-
lation of cinnamic acid, the conversion of flavanone to isoflavone and the 2'-hydroxylation of the isoflavone formononetin. Isoflavone reductase, the penultimate enzyme in isoflavonoid phytoalexin biosynthesis in alfalfa, catalyzes the formation of (R) vestitone from 2'-hydroxy formononetin (Paiva et
al., 1991). In peas, an isoflavone reductase is present which
catalyzes the formation of the corresponding 2'-hydroxy-(S)
isoflavanone (Sun et al., 1991). To date, only the caffeic
acid O-methyltransferase and (R) isoflavone reductase have
been cloned (Gowri et al., 1991; Paiva et al., 1991), but
rapid progress is now expected for the other genes. From
studies of this sort, coupled with investigations of active
site chemistry, a picture may emerge of the protein sequence
requirements and constraints determining substrate specificity, regio-specificity of substitution reactions and stereospe-
cificity in plant secondary product biosynthesis.

Enzymes of terpenoid biosynthesis - 3-Hydroxy-3-methyl-
glutaryl-coenzyme A reductase (HMGR) is the initial regulatory
enzyme for the synthesis of all isoprenoid compounds from acetate units. It has now been cloned from a number of plant
species, including Arabidopsis (Learned and Fink, 1989). Less
is known about genes encoding enzymes of specific branch pathways of isoprenoid synthesis. Some sesquiterpene cyclases
have been purified (Munck and Croteau, 1990; Vogeli et al.,
1990), and the next few years should see the cloning of their
genes and a better understanding of the catalytic mechanisms
which underlie the many different folding/cyclization patterns
of isoprenoid carbon skeletons. cDNA clones have been
obtained for the cyclase which converts geranyl pyrophosphate
to the diterpene phytoalexin casbene in castor bean (Lois and
West, 1990).

Enzymes of alkaloid biosynthesis - Many pharmacologically
important compounds are alkaloids derived from plant sources.
Economically significant alkaloids include the following:
1. codeine and morphine, analgesic phenanthrene alkaloids from
the opium poppy; 2. quinine, an antimalarial quinoline alkaloid from Chinchona bark; 3. vincristine and vinblastine, anticancer monoterpenoid indole alkaloids (mixed dimers) from
Catharanthus roseus; 4. reserpine, an anthihypertensive indole
alkaloid from *Rauvolfia serpentina*; 5. berberine, an antibiotic isoquinoline alkaloid from *Berberis* and *Coptis* species;
6. hyoscyamine or atropine, an antispasmodic tropane alkaloid
from several plants including *Hyoscyamus niger* (henbane), jim-
son weed, and deadly nightshade. Because of their commercial
importance, much research has focussed on understanding the
biosynthesis of these alkaloids, with a goal of improving
yields from plants or tissue cultures. Many enzymes have been
assayed and purified, but to date, only a few alkaloid biosynthetic genes have been cloned (see Table 3). Strictosidine
synthase was first cloned from *Rauvolfia serpentina* (Kutchan
et al., 1988), and later from *Catharanthus roseus* (McKnight *et
al.*, 1990). This enzyme catalyzes the stereospecific conden-
sation of tryptamine and a terpenoid compound to form strictosidine, a precursor of well over 1,000 known monoterpenoid
indole alkaloids. Tryptophan decarboxylase, cloned from *Catharanthus roseus* (DeLuca *et al.*, 1989) produces tryptamine, a
precursor of thousands of indole alkaloids and possibly
auxins. Hyoscyamine 6β-hydroxylase (Hashimoto *et al.*, 1990)
and (S)-tetrahydroberberine oxidase (Okada *et al.*, 1989) catalyze hydroxylation reactions very specific to hyoscyamine
(atropine) and berberine-type alkaloids, respectively.

In addition to increasing our basic knowledge of the
enzymology and regulation of alkaloid biosynthesis, the cloning of such enzymes may have practical applications, such as
increasing or altering alkaloid content. Overexpression of a
rate-limiting enzyme may increase product yields; the gene
from an already-characterized plant may be adequate. However,
the endogenous enzyme may suffer from feedback inhibition or
other undesirable catalytic features; in this case a "gene
bank" may prove to be a useful source of superior enzymes

Table 3. Cloned genes of alkaloid metabolism.

Alkaloid Class: Monoterpenoid indole
   Gene   Source

Strictosidine synthase   *Catharanthus roseus*
                                    Rauvolfia serpentina

Alkaloid Class: Indole and Isoquinoline
   Gene   Source

Tryptophan decarboxylase   *Catharanthus roseus*

Alkaloid Class: Isoquinoline (berberine)
Gene   Source

(S)-tetrahydroberberine oxidase *Coptis japonica*

Alkaloid Class: Tropane
   Gene   Source

Hyoscyamine 6β-hydroxylase   *Hyoscyamus niger*

---

through screening with heterologous probes. The two cloned strictosidine synthases show only 80% amino acid identity (McKnight *et al.*, 1990) but it is not known how this affects their activity *in vivo*. The pharmacologic properties of an alkaloid can be greatly affected by slight alterations in its chemical structure. Modified alkaloids might be produced by introducing into a cultivated plant the genes from a "donor plant" (stored DNA) which produced related alkaloids varying in methylation, hydroxylation, or stereochemistry. Such an approach has succeeded in producing novel "hybrid" antibiotics in bacteria (Hopwood *et al.*, 1985). A knowledge of the alkaloids produced by the stored (donor) plant and/or the botanical classification would be of enormous value before undertaking such a project, since the chances of success are greatest if

the donor and recipient plants make fairly similar compounds
or share common intermediates. Recognizing sequences for useful (donor) enzymes may pose a problem but, as stated earlier,
increased knowledge of enzyme structure/function relationships
may aid such efforts.

One potential drawback of the above approaches is that the
desired enzyme might not be expressed in an active form outside of its natural source. However, this does not seem to be
a common problem since tryptophan decarboxylase from *Catharanthus roseus* (Songstad *et al.*, 1990), hyoscyamine
6β-hydroxylase from *Hyoscyamus niger* (Hashimoto *et al.*, 1990)
and stilbene synthase from *Arachis hypogaea* (peanut) (Hain *et
al.*, 1990) have all been expressed in active form in tobacco
plants or callus. In fact, tryptophan decarboxylase (DeLuca
et al., 1989), isoflavone reductase (Paiva *et al.*, 1991), and
phenylalanine ammonia-lyase (Schulz *et al.*, 1989) and other
plant enzymes have been expressed in active form in *E. coli*
bacteria, aiding in positive identification of the clones. In
theory it might be possible to reconstruct an entire biosynthetic pathway in *E. coli* or tobacco for the production of
useful metabolites (Wink, 1989).

GENES ENCODING PROTEINS WITH DIRECT BIOLOGICAL ACTIVITY

A wide range of plant proteins have direct biological
activity. Enzyme inhibitors such as the proteinase inhibitors
(Ryan and An, 1988) and polygalacturonase inhibitors (De
Lorenzo *et al.*, 1990) have potential uses as protectants
against plant insect pests and pathogenic fungi respectively.
Other proteins with potential antimicrobial activity include
lectins (Broekaert *et al.*, 1989), the hydrolytic enzymes chi-
tinase and glucanase (Bowles, 1990), thionins (Bohlmann *et
al.*, 1988; Reimann-Phillip *et al.*, 1989), hydroxyproline-rich
glycoproteins (Sauer *et al.*, 1990), pathogenesis-related proteins (Walter *et al.*, 1990) and ribosome-inactivating proteins
(Barbieri *et al.*, 1989). Toxic lectins such as ricin have
great potential in cancer therapy (Vitetta *et al.*, 1983).
Sweet tasting proteins such as thaumatin (van der Wel and
Loeve, 1972), proteases such as papain and bromelain, and various amylases and glycosidases are of value in the food indus-

try.

To search for novel variants of proteins with intrinsic
biological activity through screening DNA banks with heterologous cDNAs or degenerate oligonucleotide primers is clearly
feasible, and coupled with a suitable expression system, could
lead to the identification of genes encoding proteins with
improved or altered biological activities. Even so, the
information that is lost if the only source of plant material
remaining is a DNA bank may again be that which is crucial in
allowing effective exploitation. For example, the proteinase
inhibitor I from the wild tomato species *Lycopersicon peruv-*
ianum is novel in having trypsin specificity (Wingate *et al.*,
1989). This could probably have been determined through *in*
vitro transcription/translation from a cloned source. What
would not be apparent, however, is that the inhibitor accounts
for up to 50% of the soluble protein in the fruits of this
tomato. This information immediately suggests interesting
features of the protein or its gene, such as promoter activity, mRNA stability, or protein stability which would reward
further study.

GENES ENCODING REGULATORY PROTEINS

Recent advances in our understanding of signal transduction and transcriptional regulation in mammalian systems
have led to the identification of important classes of regulatory proteins such as myb proto-oncogenes, leucine zipper and
homeobox transcription factors, G-proteins and protein
kinases. A number of plant transcription factors have now
been cloned, and homologies have been shown to myb proto-
oncogenes, leucine zipper transcription factors such as fos,
jun and GCN4, and the human serum response factor (Hartings *et*
al., 1989; Katagiri *et al.*, 1989; Marocco *et al.*, 1989; Yanofsky *et al.*, 1990). Such successes have stimulated attempts to
clone regulatory proteins from plants by homology probing or
PCR using oligonucleotides based on sequences of conserved
structural motifs. The problems of this approach are somewhat
akin to those which will be encountered on screening DNA banks
from unknown species, as the function of the cloned sequences
may not be apparent. Currently, the functional assays being

used to identify putative regulatory proteins center on studying the phenotypic effects of overexpression or underexpres-
sion of the transcript in transgenic plants. A functional
screen relying on visible phenotypes is clearly limited in its
applicability.

Homology probing has recently led to the isolation of
interesting new potential regulatory proteins. For example,
of a number of putative plant protein kinases cloned by this
strategy, some appear to be homologs of mammalian second mes-
senger protein kinases whose regulatory ligands remain to be
determined (Lawton *et al.*, 1989; Biermann *et al.*, 1990),
whereas a kinase sequence from maize appears to encode a seri-
ne-threonine transmembrane receptor kinase, whose catalytic
domain is linked through the putative transmembrane domain to
an extracellular domain similar to that of the pollen-stigma
self-incompatibility glycoprotein of *Brassica* (Walker and
Zhang, 1990). Doubtless, many novel regulatory proteins may
exist among the tens of thousands of plant species which have
not been subject to laboratory analysis. Whether the study of
such species at the DNA sequence level alone will give new
fundamental insights into plant gene regulation and signal
transduction is, however, open to debate.

CONCLUSIONS

On reviewing aspects of gene isolation, identification
and function in well investigated plant species, it becomes
clear that the utility of accessing DNA banks from endangered
species is probably not limited by technology. At present it
is the database relating protein primary sequence to function
that limits utility. While efforts to preserve DNA are com-
mendable, and will doubtless prove extremely useful for taxonomic studies, much more remains to be learned about the
biochemistry and molecular biology of common species which are
currently under investigation in laboratories throughout the
world. The sequencing of the complete *Arabidopsis* genome is a
high priority project for the next several years. This will
provide a much expanded data base for plant gene/enzyme structure, which will in turn increase the prospects for information access from gene banks of more exotic species.

Literature Cited

Barbieri, L., A. Bolognesi, P. Cenini, A. I. Falasca, A.
    Minghetti, L. Garofano, A. Guicciardi, D. Lappi, S. P.
    Miller and F. Stirpe. 1989. Ribosome-inactivating proteins
    from plant cells in culture. *Biochem. J.* 257: 801-807.
Biermann, B., E. M. Johnson and L. J. Feldman. 1990. Charact-
    erization and distribution of a maize cDNA encoding a pep-
    tide similar to the catalytic region of second messenger
    dependent protein kinases. *Plant Physiol.* 94: 1609-1615.
Bloem, L. J. and L. Yu. 1990. A time-saving method for screen-
    ing cDNA or genomic libraries. *Nucleic Acids Res.* 18:
    2830
Bohlmann, H., S. Clausen, S. Behnke, H. Giese, C. Hiller,
    U. Reimann-Philipp, G. Schrader, V. Barkholt and K. Apel.
    1988. Leaf-specific thionins of barley - a novel class of
    cell wall proteins toxic to plant-pathogenic fungi and
    possibly involved in the defense mechanism of plants.
    EMBO J. 7: 1559-1565.
Bowles, D. J. 1990. Defense-related proteins in higher plants.
    Annu. Rev. Biochem. 59: 873-907.
Broekaert, W. F., J. Van Paris, F. Leyns, H. Joos and W. J.
    Peumans. 1989. A chitin-binding lectin from stinging
    nettle rhizomes with antifungal properties. *Science* 245:
    1100-1102.
Campbell, W. H. and G. Gowri. 1990. Codon bias in higher
    plants. *Plant Physiol.* 92: 1-11.
De Lorenzo, G., Y. Ito, R. D'Ovidio, F. Cervone, P. Alber-
    sheim and A. G. Darvill. 1990. Host-pathogen interactions
    XXXVII Abilities of the polygalacturonase-inhibiting pro-
    teins from four cultivars of *Phaseolus vulgaris* to inhibit
    the endopolygalacturonases from three races of *Colletotri-
    chum lindemuthianum. Physiol. Mol. Plant Pathol.* 36:
    421-435.
DeLuca, V., C. Marineau and N. Brisson. 1989. Molecular clon-
    ing and analysis of cDNA encoding a plant tryptophan
    decarboxylase. Comparison with animal dopa decarboxy-
    lases. *Proc. Natl. Acad. Sci. USA* 86: 2582-2586.
Dixon, R. A. and M. J. Harrison. 1990. Activation, structure

and organization of genes involved in microbial defense in
plants. *Adv. Genet.* 28: 165-234.

Edington, B. V., C. J. Lamb and R. A. Dixon. 1991. cDNA cloning and characterization of a putative 1,3-β-D-glucanase
transcript induced by fungal elicitor in bean cell suspension cultures. *Plant Mol. Biol.* 16: 81-94

Edwards, K., C. L. Cramer, G. P. Bolwell, R. A. Dixon, W.
Schuch and C. J. Lamb. 1985. Rapid transient induction of
phenylalanine ammonia-lyase mRNA in elicitor-treated bean
cells. *Proc. Natl. Acad. Sci. USA* 82: 6731-6735.

_____ and R. A. Dixon. 1991a. Isoflavone 0-methyltransferase
activities in elicitor-treated cell suspension cultures
of *Medicago sativa* L. *Phytochemistry* (in press).

_____ and _____. 1991b. Purification and characterization of
S-adenosyl-L-methionine: caffeic acid 3-0-methyl-
transferase from suspension cultures of alfalfa (*Medicago
sativa* L.). *Arch. Biochem. Biophys.* (in press).

Gowri, G., R. C. Bugos, W. H. Campbell, C. A. Maxwell and
R. A. Dixon. 1991. Stress responses in alfalfa (*Medicago
sativa* L.) X. Molecular cloning and expression of S-adenosyl-L-methionine: caffeic acid 3-0-methyltransferase, a
key enzyme of lignin biosynthesis. *Plant Physiol.* (In
press).

Hahlbrock, K. and D. Scheel. 1989. Physiology and molecular
biology of phenylpropanoid metabolism. *Annu. Rev. Plant
Physiol. Plant Mol. Biol.* 40: 347-369.

Hain, R., B. Biesler, H. Kindl, G. Schrder and R. Stocker.
1990. Expression of a stilbene synthase gene in *Nicotiana
tabacum* results in synthesis of the phytoalexin resvera-
trol. *Plant Mol. Biol.* 15: 325-335.

Hartings, H., M. Maddaloni, N. Lazzaroni, N. Di Fonzo, M.
Motto, F. Salamini and R. Thompson. 1989. The 02 gene
which regulates zein deposition in maize endosperm encodes
a protein with structural homologies to transcriptional
activators. *EMBO J.* 8: 2795-2801.

Hashimoto, T., J. Matsuda, S. Okabe, Y. Amano, D. J. Yun, A.
Hayashi and Y. Yamada. 1990. Molecular cloning and tissue-
and cell-specific expression of hyoscyamine
6β-hydroxylase. Abstracts VIIth International Congress on

Plant Tissue and Cell Culture, Amsterdam, #C6-5.

Hopwood, D. A., F. Malpartida, H. M. Kieser, H. Ikeda, J. Duncan, I. Fujii, B. A. M. Rudd, H. G. Floss and S. Omura. 1985. Production of hybrid" antibiotics by genetic engineering. *Nature* 315: 642-644.

Kaiser, K. and N. E. Murray. 1985. The use of phage lambda replacement vectors in the construction of representative genomic DNA libraries. Pp. 1-47. In: D. M. Glover, (ed.) *DNA Cloning, Vol. I, A Practical Approach*, IRL Press, Oxford.

Katagiri, F., E. Lam and N-H. Chua. 1989. Two tobacco DNA-binding proteins with homology to the nuclear factor CREB. *Nature* 340: 727-730.

Kinzler, K. W. and B. Vogelstein. 1989. Whole genome PCR: application to the identification of sequences bound by gene regulatory proteins. *Nucleic Acids Res.* 17: 3645-3653.

Kutchan, T. M., N. Hampp, F. Lottspeich, K. Beyreuther and M. H. Zenk. 1988. The cDNA clone for strictosidine synthase from *Rauvolfia serpentina*. DNA sequence determination and expression in *Escherichia coli*. *FEBS Lett.* 237: 40-44.

Lawton, M. A., R. T. Yamamoto, S. K. Hanks and C. J. Lamb. 1989. Molecular cloning of plant transcripts encoding protein kinase homologs. *Proc. Natl. Acad. Sci. USA* 86: 3140-3144.

Learned, R. M. and G. R. Fink. 1989. 3-Hydroxy-3-methyl-glutaryl-coenzyme A reductase from *Arabidopsis thaliana* is structurally distinct from the yeast and animal enzymes. *Proc. Natl. Acad. Sci. USA* 86: 2779-2783.

Lindberg, R. L. P. and M. Negishi. 1989. Alteration of mouse cytochrome P450 coh substrate specificity by mutation of a single amino-acid residue. *Nature* 339: 632-634.

Lois, A. F. and C. A. West. 1990. Regulation of expression of the casbene synthetase gene during elicitation of castor bean seedlings with pectic fragments. *Arch. Biochem. Biophys.* 276: 270-277.

McKnight, T. D., C. A. Roessner, R. Devagupta, A. I. Scott and C. L. Nessler. 1990. Nucleotide sequence of a cDNA encoding the vacuolar protein strictosidine synthase from *Cath-*

aranthus roseus. *Nucleic Acids Res*. 18: 4939.

Marocco, A., M. Wissenbach, D. Becker, J. Paz-Ares, W. Saedler, F. Salamani and W. Rohde. 1989. Multiple genes are transcribed in *Hordeum vulgare* and *Zea mays* that carry the DNA binding domain of the myb oncoproteins. *Mol. Gen. Genet*. 216: 183-187.

Munck, S. L. and R. Croteau. 1990. Purification and characterization of the sesquiterpene cyclase patchoulol synthase from *Pogostemon cablin*. *Arch. Biochem. Biophys*. 282: 58-64.

Niesbach-Klosgen, U., E. Barzen, J. Bernhardt, W. Rohde, Z. Schwarz-Sommer, H. J. Reif, U. Weinand and H. Saedler. 1987. Chalcone synthase genes in plants: a tool to study evolutionary relationships. *J. Mol. Evol*. 26: 213-235.

Ochman, H., J. W. Ajioka, D. Garza and D. L. Hart. 1990. Inverse polymerase chain reaction. *Bio/technology* 8: 759-760.

Ott, R. W. and N-H. Chua. 1990. Enhancer sequences from *Arabidopsis thaliana* obtained by library transformation of *Nicotiana tabacum*. *Mol. Gen. Genet*. 223: 169-179.

Okada, N., N. Koizumi, T. Tanaka, H. Ohkubo, S. Nakanishi and Y. Yamada. 1989. Isolation, sequence and bacterial expression of a cDNA for (S)-tetrahydroberberine oxidase from cultured berberine-producing *Coptis japonica* cells. *Proc. Natl. Acad. Sci. USA* 86: 534-538.

Paiva, N. L., R. Edwards, Y. Sun, G. Hrazdina and R. A. Dixon. 1991. Molecular cloning and expression of alfalfa isoflavone reductase, a key enzyme of isoflavonoid phytoalexin biosynthesis. *Plant Mol. Biol*. (In press).

Reimann-Philipp, U., G. Schrader, E. Martinoia, V. Barkholt and K. Apel. 1989. Intracellular thionins of barley. A second group of leaf thionins closely related to but distinct from cell wall-bound thionins. *J. Biol. Chem*. 264: 8978-8984.

Ryan, C. A. and G. An. 1988. Molecular biology of wound-inducible proteinase inhibitors in plants. *Plant, Cell and Environ*. 11: 345-349.

Saiki, R. K., D. H. Gelfand, S. Stoffel, S. J. Scharf, R. Higuchi, G. T. Horn, K. B. Mullis and H. A. Erlich. 1988.

Primer-directed enzymatic amplification of DNA with a thermostable DNA polymerase. *Science* 239: 487-491.

Sauer, N., D. R. Corbin, B. Keller and C. J. Lamb. 1990. Cloning and characterization of a wound-specific hydroxy-proline-rich glycoprotein in *Phaseolus vulgaris*. *Plant, Cell and Environ.* 13: 257-266.

Schulz, W., H.-G. Eiben and K. Hahlbrock. 1989. Expression in *Escherichia coli* of catalytically active phenylalanine ammonia-lyase from parsley. *FEBS Lett.* 258: 335-338.

Somssich, I. E., J. Bollmann, K. Hahbrock, W. Kombrink and W. Schulz. 1989. Differential early activation of defense-related genes in elicitor-treated parsley cells. *Plant Mol. Biol.* 12: 227-234.

Songstad, D. D., V. DeLuca, N. Brisson, W. G. W. Kurz and C. L. Nessler. 1990. High levels of tryptamine accumulation in transgenic tobacco expressing tryptophan decarboxylase. *Plant Physiol.* 94: 1410-1413.

Sun, Y., Q. Wu, H. Van Etten and G. Hrazdina. 1991. Stereo-isomerism in plant disease resistance: induction and iso-lation of the 7,2'-dihydroxy-4', 5'-methylenedioxyisofla-vone reductase, an enzyme introducing chiralty during the synthesis of isoflavonoid phytoalexins in pea (*Pisum sativium* L.). *Arch. Biochem. Biophys.* 284: 167-173.

Timblin, C., J. Battey and W. M. Kuehl. 1990. Application of PCR technology to subtractive cDNA cloning: identification of genes expressed specifically in murine plasmacytoma cells. *Nucleic Acids Res.* 18: 1587-1593.

van der Wel, H. and K. Loeve. 1972. Isolation and character-ization of thaumatin I and II, the sweet-tasting proteins from *Thaumatococcus danielii* Benth. *Eur. J. Biochem.* 31: 221-225.

Vitetta, E. S., K. A. Krolick, M. Miyami-Inaba, W. Cushley and J. W. Uhr. 1983. Immunotoxins: a new approach to cancer therapy. *Science* 219: 644-650.

Vogeli, U., J. W. Freeman and J. Chappell. 1990. Purification and characterization of an inducible sesquiterpene cyclase from elicitor-treated tobacco cell suspension cultures. *Plant Physiol.* 93: 182-187.

Walker, J. C. and R. Zhang. 1990. Relationship of a putative

receptor protein kinase from maize to the S-locus glycoproteins of *Brassica*. *Nature* 345: 743-746.

Walter, M. H., J-W. Lin, C. Grand, C. J. Lamb and D. Hess.
1990. Bean pathogenesis-related (PR) proteins deduced
from elicitor-induced transcripts are members of uqiqui-
tous new class of conserved PR proteins including pollen
allergens. *Mol. Gen. Genet.* 222: 353-360.

Wingate, V. P. M., R. M. Broadway and C. A. Ryan. 1989.
Isolation and characterization of a novel, developmentally
regulated proteinase inhibitor I protein and cDNA from the
fruit of a wild species of tomato. *J. Biol. Chem.* 264:
17734-17738.

Wink, M. 1989. Genes of secondary metabolism: Differential
expression in plants and *in vitro* cultures and functional
expression in genetically transformed microorganisms. Pp.
239-251. In: W. G. W. Kurz (ed.), *Primary and Secondary
Metabolism of Plant Cell Cultures II*. Springer-Verlag,
New York.

Yanofsky, M. F., H. Ma, J. L. Bowman, G. N. Drews, K. A.
Feldmann and E. M. Meyerowitz. 1990. The protein encoded
by the *Arabidopsis* homeotic gene *agamous* resembles transcription factors. *Nature* 346: 35-39.

Young, D. and R. W. Davis. 1983. Yeast RNA polymerase II
genes: isolation with antibody probes. *Science* 222:
778-782.

THE DISCOVERY OF MEDICINES AND FOREST CONSERVATION

James S. Miller

Missouri Botanical Garden, 2315 Tower Grove Ave.
St. Louis, MO 63166, U.S.A.

and
Stephen J. Brewer

Monsanto Company, 700 Chesterfield Village Parkway
St. Louis, MO 63198, U.S.A.

Summary - A much greater percentage of Western medicines
owe their discovery and development to natural products
research than is generally recognized. The plant species that
comprise the forests of the world are a rich source of pharmaceuticals. In their search for new medicines, modern drug
discovery programs continue to depend on plants and microbes
as sources of chemical diversity. Programs that screen plants
as sources of biologically active chemicals have traditionally
used either random, taxonomic, or ethnobotanical collecting
strategies. Recent studies indicate that at least some animals may use plants medicinally, which could provide an additional method for locating species with high potential. Argu-
ments are presented that tropical plants are more likely to
yield biologically active compounds than the plants of temperate regions. As tropical forests are cleared, the information
lost is not only that pertaining to the individual species
of a given ecosystem but also the information concerning all
of the coevolutionary relationships among these species.

INTRODUCTION

The plants of the earth's ecosystems have proven to be
mankind's medicine chests. Eighty percent of people in developing countries rely on medicines derived from plants (Farnsworth, 1990). These people generally have an intimate knowledge of their medicinal flora and exploit a wide variety of

species; more than 1300 plant species are used medicinally in
northwestern Amazonia alone (Schultes, 1980). Drugs derived
from plants are a major part of the pharmacopia of western
medicine as well, many adopted from traditional folk remedies.
The early European botanical gardens were primarily collections of medicinal plants rather than display gardens. The
plants then formed the basis of Western medicine. The importance of natural chemicals in the development of Western medicine can be illustrated by studying the evolution of the
twenty best-selling drugs in the United States. This analysis
demonstrates that most of these modern medicines, (which
accounted for six billion dollars in sales in 1988), have
benefitted from natural products research. Plants played a
key role in the development of seven of these twenty medicines, supplying either the actual medicines, leads for medicinal chemists, or precursors for drug synthesis (Appendix A).

The top twenty selling U.S.A. drugs with total sales
approaching 6 billion dollars in 1988 (IMS International) fall
into 11 therapeutic classes (Appendix A). The role of natural
products research in the development and understanding of
these therapeutic classes and the role of screening in the
discovering the lead compound which resulted in these drugs,
was abstracted mainly from the work of Sneader (1986) but see
also Gilman et al. (1990) and Vagelos (1991). Natural products from exogenous sources (plants, microbes, and non mammalian animal tissues) were involved in understanding the pharmacology of all therapeutic classes with the exception of female
sex hormones, which were isolated from mammalian tissue. In
eight of these therapeutic classes, the natural chemicals were
used therapeutically as is, or as synthetic chemical derivatives, to treat human ailments at one stage in the development
of the current medicines. Three of the classes still use naturally derived chemicals. The role of screening in the development of these medicines was discerned by identifying how the
lead compound (the direct bioactive chemical ancestor of a
drug) was discovered. This analysis indicates that the twenty
top selling drugs were developed from fourteen lead compounds.
Random screening, where no significant structural pre-
conditions were used to select the compound, resulted in seven

of the fourteen leads. Analogue screening, where structures of a hormone or an enzyme's substrate were used to synthesize analogues, resulted in the discovery of three of the fourteen lead compounds. Unexpected observations, whether in the clinic, or during screening were responsible for the discovery of three of the fourteen lead compounds. In the case of the female sex hormones, the lead compound was the natural biochemical itself. This is in contrast with the other leads which antagonize the activity of biological molecules.

Of all of the medicines marketed in the United States, Farnsworth (1990) has estimated that 199 or about 25% contain active ingredients extracted from plants. The value of the medicines derived directly or indirectly from natural resources in 1984, was estimated at greater than 20 billion dollars (Farnsworth, 1984). The figures above indicate that directly or indirectly, the majority of Western medicines owe their existence to natural products research.

DISCOVERY OF MODERN MEDICINES

Advances in medicinal science during the last century led to the realization that specific chemical compounds are responsible for the effects of drugs. During this century research into the mode of action of these bioactive chemicals has shown that they have utility because they inhibit or stimulate specific target molecules (protein-based receptors or enzymes) in the diseased animal. In addition, the direct ancestors (leads) of today's top drugs were found by random or analogue screening, or else by chance observation (Appendix A). Therefore, the modern rational approach to drug discovery uses our knowledge of disease states and biology to identify protein based targets which when agonized or antagonized are hypothesized to yield a useful medicinal effect. However, as the structure of such bioactive chemicals cannot be adequately predicted, large numbers of chemicals, either analogs of enzyme substrates or hormones (10^2-10^3) and/or randomly selected synthetic chemicals and extracts of natural products (10^3-10^6) are screened to yield the first generation of bioactive chemicals. From the derived structure/activity relationship, a synthetic chemical analogue program can, if required,

be initiated to reduce toxicity, and improve biological activity and bioavailability. The challenge to the technologists
using the random screening technique is to obtain a large and
diverse collection of chemicals and to supply analytical technology to screen the collection in a reasonable time. Devis-
ing the screen to test large numbers of samples requires a
major commitment and considerable analytical skill. Advances
with techniques in cell biology and biotechnology also allow
most of the protein based targets to be supplied in quantities
to support large scale random screens. Microanalytical methods combined with computerized data handling allow high
throughput screens to be devised and results analyzed.
Advanced separations and structure elucidation methodologies
allow the isolation of natural product chemicals and the
assignment of structure. When combined with a keen awareness
for the opportunities available from an unexpected observation, this technology consistently produces lead compounds
suitable for development into new medicines.

PLANTS AS SOURCES OF NEW MEDICINES

Given the historical importance of plants as sources of
medicines and these improved technologies, there has been a
resurgence in interest in screening plants for pharmaceuticals
(Tyler, 1986). Three strategies exist for collecting plants
for screening programs (Cox, 1990; Rinehart et al., 1990;
Spjut and Perdue, 1976; Waterman, 1990); random, taxonomic,
and ethnobotanical. Random collecting is an attempt to sample
as much taxonomic diversity as possible. Barclay and Perdue
(1976) analyzed the results of the first 20,525 species tested
in the National Cancer Institute's plant screening program,
which relied primarily on random collection. They concluded
that genera within families often exhibited considerable chemical diversity, but species within a genus were likely to be
chemically similar. Their conclusions indicate that increasing the generic diversity among samples should increase the
chemical diversity within a given screen. One limitation of
random collecting is that it often yields samples that are
often taxonomically biased by the geographical restriction of
collecting. Thus, collecting in regions of high taxonomic

diversity is likely to increase significantly the chemical
diversity screened. Taxonomic collecting is based on the general tendency, as shown by Barclay & Perdue (1976), for
related taxa to contain related compounds. This leads to two
general applications of this method. First, when the source
of a bioactive compound is known, screening related taxa may
yield compounds of similar structure with greater efficacy or
reduced toxicity. Taxonomic screening of this type led to
examination of additional species of *Catharanthus* (Tin-Wa &
Farnsworth, 1975), after vincristine and vinblastine had been
isolated from *C. roseus* (L.) G. Don. A second application of
taxonomic collecting is to search for better sources of known
compounds. Taxonomic collecting was employed in this manner
in the search for high-yielding strains of *Cinchona*, the
source of quinine, when access to the Asian plantations became
impossible during World War II. Collecting guided by ethnobotanical data has been applied in two ways to drug discovery
programs. One approach is the study of the uses of various
plants in traditional medicine, followed by a testing of their
true effectiveness in these applications. Positive results
from this type of work depend upon careful disease recognition
and precise documentation of uses of herbal remedies, as discussed by Croom (1983). The second approach has been random
screening of plants used in traditional medicine based on the
assumption that they have a higher probability of yielding
bioactive compounds (Balick, 1990). The use of the ethnobotanical approach has the advantage of providing a "prescreen"
in the sense of Cox (1990). Another potential "prescreen" for
pharmacologically active plants, not previously discussed in
the literature, is based on observations that some animals may
use certain plants as medicines in the same manner as indigenous peoples. For example, chimpanzees in eastern Africa
apparently utilize plants to treat various ailments. The animals generally chew the leaves of their food plants but swal-
low whole the leaves of three species of *Aspilia* Thours.
(Asteraceae) (Wrangham & Nishida, 1983), which have been found
to be rich in the bioactive compound Thiarubine-A (Rodriguez
et al., 1985). Chimpanzees have also been reported to use
Vernonia amygdalina Del. (Asteraceae) to combat symptoms of

lethargy, lack of appetite, and irregularity of bodily excre-
tions (Huffman & Seifu, 1989). The same species of *Vernonia*
is used by local people to treat a wide variety of medical
problems. As with *Aspila*, the chimpanzees ate *Vernonia* in an
unusual manner. The pith was chewed out of stems, the astrin-
gent juice was swallowed, and the remains were spit out. An
analogous situation was reported for elephants in east Africa
in which a tree of the family Boraginaceae was eaten to induce
labor (Cowen, 1991). Again, the same plant species is used
for the same medical function by local women. As studies of
this type emerge, they may provide another means of identifying plants likely to yield useful pharmaceuticals.

IMPORTANCE OF TROPICAL FORESTS

Tropical forests are much more diverse than temperate forests and a much smaller proportion of the species in tropical
areas have been studied, either in general or as possible
sources of new drugs. While 0.1 ha. plots of forest in the
central United States contained 20-26 species of woody plants
with a diameter of 2.5 cm diameter at breast height, plots of
similar size in the forests of western Colombia yielded
258-265 species (Gentry, 1988), approximately a tenfold
increase in the number of species per unit area. There is
also good reason to believe that individual tropical species
are more likely to yield bioactive compounds than those from
temperate regions. Most bioactive compounds derived from
plants are secondary metabolites that have the presumed function of protecting the plant from herbivores or pathogens.
Plant secondary compounds have been classified as qualitative,
compounds with presumed specific toxicity, or quantitative,
compounds that reduce the digestibility of the plant (Hegarty,
in press). Hartwell (1976) surveyed the chemical compounds
exhibiting antitumor activity that had been isolated from
higher plants in the National Cancer Institute's drug discovery program. Of the twelve classes he examined, the quassi-
noids, ansamacrolides, and alkaloids were regarded as the most
promising and all are qualitative secondary compounds. Most
compounds of this type are thought to be defenses that plants
produce against bacterial or fungal infections or herbivory

and the evolutionary selection pressures that lead to production of these qualitative compounds should increase in proportion to the diversity of insects and pathogens that confront a
plant population. This in turn suggests that the production
of these compounds should be highest in the tropics where
organismal diversity is greatest. Levin (1976) showed that
tropical plants were richer in alkaloids than those from temperate regions, and that overall alkaloid diversity was
inversely correlated with latitude. He concluded that this
was a response to greater diversity of herbivores in the tropics where selection favors a greater variety of defense compounds. Microbial diversity is also much greater in tropical
forests, so the same pattern should also hold for phytoalexins, a group of compounds that are produced transiently in
response to microbial infection and have shown promise as
potential pharmaceuticals (Barz et al., 1990). Another reason
for intensifying work in tropical regions is their threatened
environmental status. The tropical forests, are also the most
diverse with about 40% of the world's organisms (Myers, 1984)
in only 7% of the land surface area. The temperate ecosystems
of North America and Eurasia have been seriously altered by
man, but have achieved some stability and, although they are
still under human pressure, less extinction due to human activity is expected than in tropical areas. If the extremely
high extinction rates predicted for the tropics (Simberloff,
1986) are accurate, priority should be given to sampling tropical species while they still exist.

CONSERVATION OF TROPICAL FORESTS

The forests of the world can continue to be an important
source of medicines. Those of tropical regions are of particular interest in that they have the highest taxonomic diversity and in addition tropical plants may be more diverse in
qualitative secondary metabolites. Yet the forests with the
most potential are rapidly vanishing. The tropical forests
that originally covered 16% of the earth's land surface have
been reduced today to 7% of its surface (Gentry, 1990). Madagascar retains less than 10% of its original forest cover, and
regions with large forest tracts that are still intact, such

as Amazonia, are under heavy threat. It has been predicted
that these large forest expanses will be reduced to small forest patches early in the next century (Simberloff, 1986). As
these forests with their component species disappear, a valuable source of potential pharmaceuticals and other natural
products will disappear with them. As Cox (1990) and Balick
(1990) have argued, ethnobotany may provide us with a more
rapid method of assessing the pharmaceutical value of individual species when there is not time to test them all. The
studies cited above indicate that we may be able to derive
similar types of information from animals. However, these
studies also tell us something else; that an intact ecosystem
with its full complement of plant and animal species, as well
as indigenous peoples, contains far more information than the
sum of its component species. The fungi that can exist, both
free-living and symbiotically are being lost rapidly and
irreversibly as cutting and clearing progresses. Many of
these live only in intact forests, and have been a rich source
of medicines. As the tropical forests are reduced to patches
smaller than a minimum critical survival size, component
species are lost, and consequently we lose not only the information in the species themselves but also the valuable information concerning their coevolutionary relationships, interre-
lationships, and interactions. While seed banks and DNA banks
allow us to preserve some of this information, and are probably the only viable means for saving many of the tropical
species threatened with extinction, we can preserve the full
information content of a forest only if all of its component
species are maintained in their original setting.

Acknowledgements
      We thank Charlotte Taylor for comments on the manuscript.

Literature Cited

Balick, M. J. 1990. Ethnobotany and the identification of therapeutic agents from the rainforest. Pp. 22-39. In: D. J. Chadwick and J. Marsh (eds.), *Bioactive compounds from plants*. Wiley, Chester. Ciba Foundation Symposium 154

Barclay, A. S. and R. E. Perdue Jr. 1976. Distribution of anticancer activity in higher plants. *Cancer Treatment Reports* 60 (8): 1081-1113.

Barz, W., W. Bless, G. Borger-Papendorf, W. Gunia, U. Mackenbrock, D. Meier, Ch. Otto, and E. Super. 1990. Phytoalexins as part of induced defence reactions in plants: their elicitation, function and metabolism. Pp. 140-156. In: D. J. Chadwick and J. Marsh (eds.), *Bioactive compounds from plants*. Wiley, Chester. Ciba Foundation Symposium 154

Cowen, R. 1991. Medicine on the wild side. *Science News* 138: 280-282.

Cox, P. A. 1990. Ethnopharmacology and the search for new drugs. Pp. 40-47. In: D. J. Chadwick and J. Marsh (eds.), *Bioactive compounds from plants*. Wiley, Chester. Ciba Foundation Symposium 154.

Croom, E. M. 1983. Documenting and evaluating herbal remedies. *Econ. Bot.* 37: 13-27.

Farnsworth, N. R. 1984. How can the well be dry when it is filled with water? *Econ. Bot.* 38: 4-13.

_____. 1990. The role of ethnopharmacology in drug development. Pp. 2-11. In: D. J. Chadwick and J. Marsh (eds.), *Bioactive compounds from plants*. Wiley, Chester. Ciba Foundation Symposium 154.

Gentry, A. H. 1988. Changes in plant community diversity and floristic composition on environmental and geographical gradients. *Ann. Missouri Bot. Gard.* 75: 1-34.

_____. 1990. Tropical forests. Pp. 33-43. In: A. Keast (ed.), *Biogeography and ecology of forest bird communities*. Academic Publishing, The Hague, The Netherlands.

Gilman, A. G., T. W. Rall, A. S. Nies, and P. Taylor. (eds.). 1990. *Goodman and Gilman's: The pharmacological basis of therapeutics*. Eighth Edition. Pergamon Press, New York.

Hartwell, J. L. 1976. Types of anticancer agents isolated from plants. *Cancer Treatment Reports* 60 (8): 1031-1067.

Hegarty, M. P. 1991. Secondary chemistry of vines. In: F. E. Putz and H. Mooney (eds.), *Biology of climbing plants.* Cambridge Univ. Press, Cambridge. (in press).

Huffman, M. A. and M. Seifu. 1989. Observations on the illness and consumption of a possibly medicinal plant *Vernonia amygdalina* (Del.), by a wild chimpanzee in the Mahale Mountains National Park, Tanzania. *Primates* 30: 51-63.

Levin, D. A. 1976. Alkaloid-bearing plants in ecogeographic perspective. *Amer. Nat.* 110: 261-284.

Myers, N. 1984. *The primary source.* W. W. Norton & Company, New York.

Rinehart, K. L., T. G. Holt, N. L. Fregeau, P. A. Keifer, G. R. Wilson, T. J. Perun Jr., R. Sakai, A. G. Thompson, J. G. Stroh, L. S. Shield, D. S. Siegler, L. H. Li, D. G. Martin, C. J. P. Grimmelikhuijzen, and G. Gade. 1990. Bioactive compounds from aquatic and terrestrial sources. *J. Nat. Prod.* 53: 771-792.

Rodriguez, E., M. Aregullin, T. Nishida, S. Uehara, R. W. Wrangham, Z. Abramowski, A. Finlayson, and G. H. N. Towers, 1985. Thiarubine-A, a bioactive constituent of *Aspilia* (Asteraceae) consumed by wild chimpanzees. *Experientia* 41: 419-420.

Schultes, R. E. 1980. The Amazonia as a source of new economic plants. *Econ. Bot.* 33: 259-266.

Simberloff, D. S. 1986. Are we on the verge of a mass extinction in tropical rain forests? Pp. 165-180. In: D. K. Elliot (ed.), *Dynamics of extinction.* Wiley & Sons, N.Y.

Sneader, W. 1986. *Drug discovery: The evolution of modern medicines.* John Wiley & Sons, New York.

Spjut, R. W. & R. E. Perdue Jr. 1976. Plant folklore: a tool for predicting sources of antitumor activity? *Cancer Treatment Reports* 60(8): 979-985.

Tin-Wa, M. & N. R. Farnsworth. 1975. The phytochemistry of minor *Catharanthus* species. Pp. 85-124. In: W. I. Taylor and N. R. Farnsworth (eds.), *The Catharanthus alkaloids.* Marcel Dekker, New York.

Tyler, V. E. 1986. Plant drugs in the twenty-first century. *Econ. Bot.* 40: 279-288.

Vagelos, P. R. 1991. Are prescription drug prices high? *Science* 252: 1080-1084.

Waterman, P. G. 1990. Searching for bioactive compounds: various strategies. *J. Nat. Prod.* 53: 13-22.

Wrangham, R. W. & T. Nishida. 1983. *Aspila* spp. leaves: a puzzle of feeding behavior of wild chimpanzees. *Primates* 24: 276-282.

Appendix A. Roles of natural products and screening in the
development of the top selling U.S. medicines.

USE/PRODUCT NAME: Adrenaline Antagonists/Atenolol & Metoprolol

ROLE OF NATURAL PRODUCTS: Ergot, a fungus which infected and
poisoned wheat had been used medicinally since the Middle
Ages. One isolated active alkaloid was ergotoxine. This had
the new property of reversing (antagonizing) many of the
actions of adrenaline, notably that of increasing blood pressure (1905).

ROLE OF SCREENING: In a screen for bronchodilators dichlo-
roisoprenaline was observed, to block adrenaline action
(c1957). It was recognized that this unexpected activity
could protect the hearts of patients with coronary disease
from adrenaline released by physical or emotional stress.
Medicinal chemistry of this lead increased the selectivity for
heart receptors and reduced toxicity.

USE/PRODUCT NAME: Anti-inflammatory/Piroxicam, Sulinac and
Naproxen

ROLE OF NATURAL PRODUCTS: A glucose derivative of salicylic
acid (salicin) is produced by willows. These grow in cold
damp places which is also conducive to rheumatic fever.
Influenced by herbalist doctrines that antidotes are found in
the vicinity of poisons, salicylic acid was tested in patients
as an antirheumatic, and shown to be beneficial (1874).

ROLE OF SCREENING: Piroxicam was the product of extensive
chemical programs to overcome the toxicity of salicylic acid.
Other leads were found by screening in an animal inflammatory
model. Sulinac was developed from an idomethacin lead discovered by screening serotonin analogs (c1963) and the phenoxyal-
kanoic acid lead for Naproxen was found (c1962) by random
screen of chemicals (originally a herbicide). Medicinal chemistry reduced toxic effects.

USE/PRODUCT NAME: Antianginals/Diltiazem & Nifedipine

ROLE OF NATURAL PRODUCTS: The first antianginals were organo
nitriles of glycerol and amyl alcohol which were shown to
dilate blood vessels (1865). Clinical tests showed that these
compounds could treat angina. Glycerol and amyl alcohol were
both natural products isolated respectively from the manufac-
ture of soap, from animal fats and alcohol, from yeast fermentation.

ROLE OF SCREENING: In a routine animal screen, analogs of
dihydropyridine were shown to cause a relatively long lasting
drop in the blood pressure of dogs (1965). This was shown to
be due to coronary vasodilating activity. An extensive medicinal chemistry program improved the oral activity of this
lead.

USE/PRODUCT NAME: Antianxiety/Alprazolam

ROLE OF NATURAL PRODUCTS: Reserpine, an alkaloid isolated from
snake root (an Indian herbal medicine able to "calm violent
lunatics") helped define the pharmacology of tranquilizers
(c1944). The first antianxiety drug was described in 1946.
This was unexpectedly found when testing a synthetic chemical
developed as an alternative to tubocurarine (a neuromuscular
blocking alkaloid).

ROLE OF SCREENING: A benzodiazepine lead was discovered by
randomly screening chemicals for muscle relaxant and antianxiety effects in behaviorally suppressed rats (1958). Medicinal chemistry resulted in Alprazolam, a benzodiazepine with
improved metabolism and reduced toxicity.

USE/PRODUCT NAME: Antibiotics/Cefaclor, Ceftriaxone &
Cefoxitin

ROLE OF NATURAL PRODUCTS: A semisynthetic quinine derivative,
ethylhydrocupreine was the first antibacterial chemotherapeu-
tic agent introduced into medicine (c1911). Although anti-

biotic properties of microbial extracts were recognized since
the 19th century, penicillin was the first product used in a
pure form (1941). Microbial fermentations are currently used
to produce precursors of these.

ROLE OF SCREENING: These antibiotic products were derived from
the beta-lactam antibiotic, cephalosporin. Cephalosporins
were discovered as part of a microbial screening program. The
organism was isolated from a sample obtained from a sewer out-
let (1948). Medicinal chemistry resulted in an increased
spectrum of antibacterial activity and the production of an
orally active drug.

USE/PRODUCT NAME: Antihistamine/Terphenadine (H1) Ranitidine &
Cimetidine (H2)

ROLE OF NATURAL PRODUCTS: Bacterial decomposition of ergot (a
medicinal fungus which grows on wheat) gave a product which
stimulated uterine contractions. Histamine was identified as
the active principle (1910). Later it was proved to be present in animal tissue (c1926). This stimulated the search for
an antagonist to overcome the toxic effect of histamine
(1885).

ROLE OF SCREENING: Terphenadine resulted from extensive medicinal chemistry of piperoxan, a lead found by a random screen
of chemicals which protect animals from histamine toxicity
(c1937). The H1 class treat allergies and motion sickness.
Leads for the H2 antihistamines were discovered by screening
histamine analogs for gastric acid inhibition in rats (c1968).
Medicinal chemists improved specificity and oral activity.
They are used to treat gastric ulcers.

USE/PRODUCT NAME: Antihypertensives/Captopril & Enalapril

ROLE OF NATURAL PRODUCTS: The ability of a snake venom to
drastically reduce blood pressure was shown to be caused by
its inhibition of angiotensin converting enzyme (1968). This
enzyme produces an octapeptide which raises blood pressure and

also destroys bradykinin which decreases blood pressure. Thus
it became evident that an inhibitor of this enzyme could be an
effective antihypertensive.

ROLE OF SCREENING: Angiotensin converting enzyme resembles
carboxypeptidase A which was known to be inhibited by 2-ben-
zylsuccinic acid. Based on this lead and on models of the
enzyme active site, potential inhibitors were synthesized and
screened for enzyme inhibition (1977). Medicinal chemistry
improved the oral activity of these inhibitory peptide analogs.

USE/PRODUCT NAME: Antipyretic/Paracetamol

ROLE OF NATURAL PRODUCTS: Quinine, isolated from cinchona bark
and salicylic acid, originally isolated from *Spirea ulmaria*,
were the first drugs used as anitipyretics. Acetylsalicylic
acid (aspirin) was synthesized to reduce unpleasant taste
(c1898). Attempts to improve quinine resulted in the production of the synthetic quinoline antipyretics (1885).

ROLE OF SCREENING: The lead compound was acetanilide whose
antipyretic effects were discovered by chance when this material was incorrectly administered in place of naphthalene,
which was being tested for the treatment of intestinal worms
(1886). Paracetamol, an analog with reduced toxicity, was
prepared in 1893 but due to misleading toxicity date, was not
marketed until 1953.

USE/PRODUCT NAME: Diuretic/Hydrochlorothiazine

ROLE OF NATURAL PRODUCTS: Numerous herbs were known to induce
mild diuresis. The xanthine alkaloids from tea, cocoa and
coffee were identified as active chemicals. Theophylline,
isolated from *Camellia* (tea) was three times as potent as caf-
feine and clinically used as a diuretic (1902) until replaced
by more potent synthetic drugs.

ROLE OF SCREENING: Carbonic anhydrase was shown to be impor-

tant in water resorption from the kidney because its inhibition by antibacterial sulphonamides caused alkaline diuresis
(1942). A chemical which had undergone an unexpected ring
closure to form a benzothiadiazine was tested in a random
screen for inhibitors of this enzyme (c1955). Medicinal chemistry increased the potency of this lead.

USE/PRODUCT NAME: Female sex hormones/Estrogenic Substances

ROLE OF NATURAL PRODUCTS: The first attempts at hormone therapy used extracts of animal tissue (c1912). Some estrogenic
compounds are found in plants but these are not medicinally
important. However, plants are important commercial sources
of synthetic precursors of these hormones.

ROLE OF SCREENING: Female sex hormones were isolated from
ovarian extracts of animals (1929) using an assay based on
changes in cells lining the vaginal walls of immature mice
which paralleled those seen in the menstrual cycle of mature
mice. Female sex hormones with reduced toxicity and improved
activity have been developed by screening analogs of natural
hormones.

USE/PRODUCT NAME: Hypercholesterolemia/Lovastatin

ROLE OF NATURAL PRODUCTS: Hydroxymethyl-CoA-reductase is an
enzyme involved in the biosynthesis of cholesterol (c1958). It
was rationalized that inhibition of this enzyme would reduce
the levels of cholesterol which were involved in the development of atherosclerosis. Lovastatin, discovered in 1979 is a
napthalenyl ester inhibitor produced by fungal fermentation.

ROLE OF SCREENING: This fungal metabolite was discovered by a
mechanism based natural product screen. Extracts of diverse
microbes are tested for inhibition of hydroxymethyl-CoA-
reductase, an enzyme involved in the biosynthesis of cholesterol (1978). This product demonstrates the role of natural
products in modern drug discovery programs.

PRESERVATION OF DNA IN PLANT SPECIMENS
FROM TROPICAL SPECIES BY DESICCATION

Robert P. Adams and Nhan Do

Plant Biotechnology Center, Baylor University,
BU Box 97372   Waco, TX 76798-7372 U.S.A.

and
Chu Ge-lin

Institute of Botany, Northwest Normal University
Lanzhou, Gansu, China

Summary - The preservation of genomic DNA in spinach
leaves by desiccation, using drierite and silica gel, was
examined. Drierite and silica gel were found to be equally
effective in preserving DNA in spinach leaves for up to 6
months at 37°C. Similar results were obtained using *Juniperus*
leaves but *Magnolia* leaves had more degradation and oak (*Quercus*) leaves had considerable degradation after 6 months storage in silica gel at 37°C. Storage tests involving fresh and
dried spinach leaves indicated that either materials can be
used, but air dried spinach leaves had slightly less DNA
degradation at each sample interval. The use of either silica
gel or drierite appears to be an acceptable method for
the preservation of DNA in tropical plant leaf specimens for
subsequent extraction and archiving or analyses. A practical
field protocol for DNA collection is presented.

INTRODUCTION

It is now estimated that the principal areas of diversity
among plants, the lowland tropical forests, will have been cut
over or severely damaged within the next 20 - 30 years (Raven,
1987, 1988). The diversity in these areas is immense: the
Amazon River system, for example, contains eight times as many
species as the Mississippi River system (Shulman, 1986). Raven
(1987, 1988) estimates that approximately 60,000 plant species

(1 of 4) will go extinct with the loss of the tropical rain
forests and as many as 1.2 million species (plants and animals) will go extinct in the next twenty to thirty years without factoring in the long-term losses due to global greenhouse
effects. The extinctions of plant species mean the loss of
thousands of links in plant phylogeny which will seriously
hinder phylogenetic and evolutionary studies forever. The
extinctions also mean a loss of valuable genes that could be
put into our current and potential crop species by genetic
engineering.

Relatively few scientists were interested in a 'genetic
insurance policy' when the idea of banking genomic DNA from
plants was first proposed (Adams, 1988). However, currently
there are 30 institutions (representing 21 nations and every
continent) in DNA Bank-Net (Figure 1).

DNA Bank-Net

Figure 1. Map of cooperating and projected members of DNA
Bank-Net. Additional nodes in the DNA Bank-Net will be added
as the association develops.

Collections of plant specimens have been utilized for the formulation of our understanding of morphological variation among taxa. Indeed, without the great herbaria of the world, our knowledge of plant evolution would be fragmented at the least. As we have moved into the era of utilizing chemical data for systematic and evolutionary studies, methods of pre-serving plant materials for future (chemical) work have been largely ignored. We are usually content to file a voucher specimen to document our chemical studies. With the present level of support for plant collections it is unlikely that much of the world's plant species can be preserved by freezing so that scientists might have access to the study of secondary compounds, enzymes, or DNA/RNA in the coming centuries.

The problems associated with bringing back fresh or frozen materials can generally be overcome by specialists (ex. world-wide collections of fresh foliage of *Juniperus* for essential oil analyses and DNA by RPA). However, botanists doing flor-istic research will collect the bulk of the specimens from tropical rare and endangered species. They often collect specimens from scores of different species in a single day. The bulk of the materials that they have to process and ship requires that any protocol for the collection of samples for specialized needs (ex. DNA storage/analyses), must be quick, simple and trouble-free. The generalist collector, working in tropical areas cannot be expected to preserve hundreds or thousands of collections for months under tropical conditions and then arrange transport through customs, all the while keeping the individual specimens frozen.

Fortunately, at least as far as DNA preservation is con-cerned, the ancient Egyptians provided a clue to the solution. Paabo (1985) reported that cloned DNA from dehydrated mummy tissues (2,400 y old) "seemed to contain little or no modifi-cations introduced postmortem". He further postulated that the use of crystalline salts for desiccation prevented the normal hydrolytic processes (DNase activities). DNAs from the more readily dehydrated tissues (skin, outer ear) were much better preserved than tissues from internal organs that were more difficult to dehydrate (Paabo, 1985). DNA has also been obtained from the extinct quagga (dried muscle attached to

salt-preserved skin, Higuchi *et al.*, 1984), a 5000 year old
Egyptian mummy (Paabo, 1989) and a 7000 year old brain (Paabo
et al., 1988) preserved in a peat bog.

Only a few papers have been published dealing with the
preservation/degradation of DNA in plant specimens. Rogers
and Bendich (1985) obtained DNA with a maximum length of 20 to
30 kbp with average lengths of 0.1 to 1.0 kbp from herbarium
vouchers ranging in age from 20 to 95 years old. DNA from
Juniperus osteosperma seeds from pack rat middens (3,500;
27,000; and greater than 45,000 y bp) yielded DNA with a maximum lengths of 10, 10, and 10 kbp and average lengths of 7, 5,
and 3 kbp, respectively.

Doyle and Dickson (1987) attempted to preserve *Solanum*
leaves using formalin-acetic acid-ethanol (FAA), Carnoy's
solution (ethanol:acetic acid, 3:1), 70% ethanol, chloro-
form:ethanol (4:3), brine solution (10% NaCl), and drying at
42°C. Although none of the chemical treatments preserved DNA
for more than a few weeks, Doyle and Dickson (1987) did obtain
adequate sized DNA from air dried leaves.

Dally and Second (1989) reported that dry leaf powder
could be stored for many months over P_2O_5 (a very powerful
desiccant) at -20°C. Unfortunately, no data was presented
about storage at higher temperatures as encountered when shipping materials by surface freight both within and between
tropical countries. Nor did Dally and Second (1989) comment
on the use of a safer desiccant than the highly corrosive
P_2O_5.

Pyle and Adams (1989) examined the use of acetic
acid/ethanol (1:3), ethanol (95%), glycerol, Na azide (3%),
Perfix preservative, pentachlorophenol/chloroform (1.5/98.5),
paraformaldehyde (10%), guanidine thiocyanate (1%), NaCl
(25%), Chlorox (5%), methanol/chloroform/propionic acid
(1:1:1), glutaraldehyde (8%), formaldehyde (7.4%), trichlo-
roacetic acid, (10%), glutaraldehyde (2%)/0.5M Na cacodylate
and 10mM EDTA to preserve spinach leaves. None of the aforementioned solutions preserved the DNA in spinach for 7 days
(Pyle and Adams, 1989). However, the preservation of DNA by
desiccating leaves in silica gel or drierite was reported by
Pyle and Adams (1989). They obtained high molecular weight

DNA from spinach following desiccation and storage over drierite in a desiccator after 2 months but the DNA was degraded
when tested at 5 months (see also Adams, 1990).

Liston, *et al*. (1990) reported good DNA yields from numerous specimens from Xinjiang, China that had been field preserved in drierite. Chase and Hills (1991) used silica gel to
preserve leaves of *Ericaceae* and *Malpighiaceae* species with
success. Both of these studies analyzed the DNA after only a
few weeks in the desiccant. No long term studies have been
reported to date.

Those results prompted us to perform additional storage
tests in our lab using both fresh and air dried leaves of spinach, juniper, magnolia, and oak by placing the leaves
directly in contact with silica gel and drierite in jars. In
addition, both drierite and silica gel were further investigated in regards to drying efficiency and capacity.

PREPARATION OF PLANT MATERIAL

Leaves from *Juniperus virginiana* L. and *Quercus virginiana*
Mill. were collected from native trees on the Baylor University campus. *Magnolia grandiflora* L. leaves were collected
from cultivated trees on the Baylor campus. Leaves from fresh
spinach (*Spinacia oleracea* L.) were purchased locally.
Approximately 0.5 g (fw) of leaf was used for each storage
treatment. Leaf drying was done in a conventional plant press
dryer at $42^{o}C$ for 24 - 48 h. A few indicating crystals were
mixed the drierite (anhydrous $CaSO_4$, W. A. Hammond Drierite
Co.) or silica gel prior to drying. This allows visual
inspection of the jars in storage as the blue crystals turn
pink when hydrated. A jar with a cracked or loose cap, for
example, can be easily detected. Desiccation was performed by
half filling a 30 ml screw capped glass jar with drierite or
silica gel, adding the leaf, and then completely filling the
jar with the appropriate drying agent. The jars were stored
in an incubator at $37^{o}C$ until analyzed (in triplicate).

For the experiments on the determination of the drying
rates for drierite and silica gel, fresh spinach leaves were
cut into 2 cm x 2 cm squares (0.5 g FW) and placed in jars
with 20 g of drierite or silica gel at $22^{o}C$ and sealed. After

24, 48, 72, 96 or 120 h, the leaves were taken from 3 jars of
both drierite and silica gel, weighed, then dried in a microwave oven for 8 min (700 watts) and weighed immediately. Preliminary experiments showed that 4 min at 700 watts in the
microwave was equivalent to oven drying at 100°C for 48 h. To
be extra careful, the leaves were dried an additional 4 min (8
min total) in the microwave on high (700 watts). Percent dry
matter was calculated as: 100 x dry weight / weight before
drying.

DNA EXTRACTION AND ANALYSES

The hot CTAB procedure (Doyle and Dickson, 1987; Doyle and
Doyle, 1987) was used for DNA extraction with minor changes as
noted in Pyle and Adams (1989). Studies (Birren et al., 1989;
Bostock, 1988; Carle, Frank and Olson, 1986; Gekeler et al.,
1989; Devos and Vereruysse-Dewitte, 1989) on the separation of
high MW DNA (up to several hundred kbp) by field inversion gel
electrophoresis (FIGE) prompted us to use FIGE for the analyses of genomic DNA in this study. The extracted DNAs were
run on a 1% agarose gel, using a BioRad mini-sub electrophoresis unit, and a BioRad Pulsewave 760 to invert the current (70
v), starting at 3 sec forward and 1 sec reverse, ramped to 12
sec forward and 4 sec reverse at the end of 4 h. T5 DNA
(Sigma D8010, 103 kbp), Lambda DNA (Sigma D9768, 48.5 kbp) and
HindIII restricted lambda DNA (23.13, 9.614, 6.557, 4.361,
2.322 and 2.027 fragments used) were co-run to provide size
standards. Two µl of a 20 ng/µl fluorescein isothiocynate
dextran (Sigma FD2000S, 2 million MW) solution was added to
each well just prior to photographing in order to clearly
delineate the front of the wells on the densitometer scans.
The Gels were photographed under short wave UV light using a
Polaroid direct screen camera (DS34). DNA was quantified by
use of a video densitometer (JAVA video analysis system, Jan-
del Scientific Inc.; CCD video camera, model WV-BL200, Pan-
asonic Corp.; PC Vision frame grabber, Image Technology,
Inc.). Average and mode molecular weights and the percent (%)
of the DNA greater than 6 kbp were calculated using program
GRAFGEL (program available for IBM PC from RPA). Restricti-
bility of stored DNA was checked using EcoRI, Hind III, Kpn I,

and Sal I (Sigma R2132, R1882, R9506 and R0754, respectively).

RESULTS AND DISCUSSION

    The storage of fresh spinach leaves (figure 2) in silica gel for up to six months appears to be satisfactory. Notice

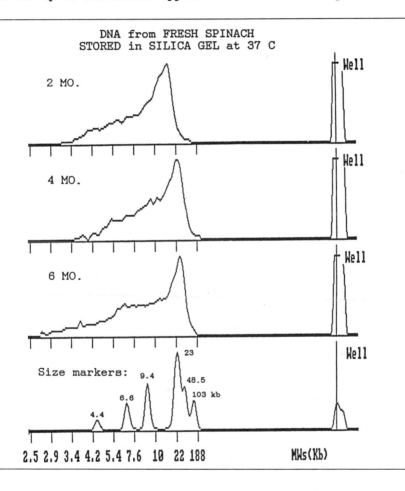

Figure 2. Video densitometer scans of FIGE of genomic DNA from fresh spinach leaves stored in silica gel at 37°C for 2, 4 and 6 months. Notice the gradual increase in lower molecular weight DNA (the peak tailing to the left). Size markers are : lambda DNA cut with HindIII (2.3, 2.0 and 0.5 kbp bands not shown); Lambda DNA (48.5 kbp) and T5 DNA (103 kbp).

that the genomic DNA ranges from about 190 kbp down to about 2
kbp (after 6 mos., Figure 2). The mode and average MWs (in
kbp) were: 2 mo.- 13.4, 10.7; 4 mo.- 24.5, 18.6; and 6 mo.
-29.7, 16.5. The only explanation we have for the increasing
mode and average MWs with time in storage is that several

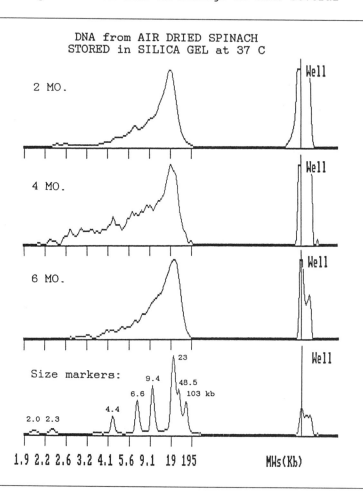

Figure 3. Video densitometer scans of FIGE of DNA from air
dried spinach leaves stored in silica gel at 37°C for 2, 4 and
6 months. Size markers are: lambda DNA cut with HindIII (0.5
kbp band not shown); Lambda DNA (48.5 kbp) and T5 DNA (103
kbp).

changes in the protocols were made (such as cutting off the
Eppendorf pipet tips before transferring DNA) during the span
of this research. It seems likely that the increases in the
mode and average molecular weights were due to improvements in
techniques during the experiment.

The DNA in the air dried spinach leaves appears similar to
DNA from fresh leaves (Figure 3). The mode and average MWs
increased in the 6 mo. sample (mode, average MW in kbp: 2 mo.
- 15.2, 15.1; 4 mo. - 15.2, 12.2; 6 mo. - 19.2, 16.1) just as
with the fresh spinach. One might note that the densitometer
tracing for the 4 mo. sample (Figure 3, second scan) displays
a rather jagged trace from 2 to 10 kbp. This due to the
sample being under-loaded on the gel (i.e., high background
noise to signal ratio). The trace for the 6 mo. sample has
much less noise due to a higher sample loading level (DNA was
more concentrated in the sample). Comparisons of three different storage tests are shown in Figure 4. None of the
stored leaves had as much high molecular weight DNA as fresh
spinach, but the DNA from air dried leaves, stored in silica
gel at 22^0 for 8 months, was quite comparable. The lower storage temperature (22^0 vs. 37^oC) appears to make a difference
in the average molecular weight.

Analyses of the effects of two storage temperatures (22^oC
and -20^oC) on the DNA in air dried spinach leaves stored in
silica gel or paper envelopes gave the following results
(after 8 mo. storage):

Treatment	Avg. MW(kbp)	Mode MW(kbp)
silica gel, -20^oC	31.5	27.7
silica gel, 22^oC	28.6 kbp	33.2
envelope, -20^oC	34.0	29.2
envelope, 22^oC (1 mo.)	< 1 kbp	< 1 kbp

It should be noted that during the first month of storage, we
experienced almost continual rainfall for two weeks and the
ambient relative humidity in the laboratory was very high (90
-100%). Apparently the paper envelopes absorbed ambient
moisture and rehydrated the leaves, leading to DNA breakdown.

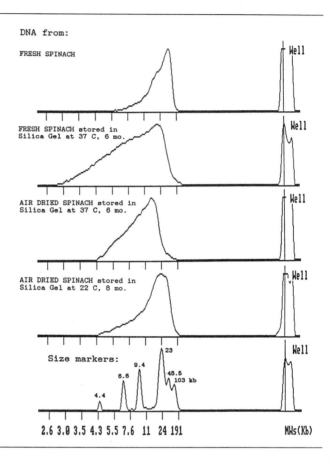

Figure 4. Video densitometer scans of FIGE of genomic DNA
from: fresh spinach leaves; fresh spinach leaves stored in
silica gel at 37°C for 6 months; air dried spinach leaves
stored in silica gel at 37°C for 6 months; and air dried spinach leaves stored in silica gel at 22°C for 8 months. Size
markers are: lambda DNA cut with HindIII (2.3, 2.0 and 0.5 kbp
bands not shown); Lambda DNA (48.5 kbp) and T5 DNA (103 kbp).

There were no significance differences between any of the
average and mode MWs for the silica gel, -20°/22°C or envelopes, -20°C treatments. However, due to the small number of
replicates (3) per treatment, one must view these results as

preliminary.

Results similar to spinach were obtained using fresh leaves of *Juniperus virginiana*. However, the DNA from magnolia (*Magnolia grandiflora*) leaves showed considerably more degradation after 6 months at 37°C than spinach and the DNA from liveoak (*Quercus virginiana*) was even more degraded. The problems with the storage of magnolia and liveoak leaves are not understood at present and additional research is being conducted. It is possible that interfering cellular components (such as polysaccharides, see Do and Adams, 1991) form complexes or lipids produce free radicals that degrade the DNA during storage and/or extraction. The thick waxy cuticles on liveoak and magnolia may have slowed the escape of moisture, leading to conditions favoring DNA degradation. In any case, avoiding high storage temperatures is beneficial and one should place the materials into a freezer as soon as they are received.

Generally, the DNA from desiccated fresh spinach shows more small (2 - 6 kbp) DNA fragments than the DNA from desiccated air dried spinach leaves. We can only speculate that some DNase activity occurs in the fresh leaves before the silica gel effectively dehydrates the leaves and denatures the DNases. Labuza (1970) reviews the properties of water in food preservation and states that 'water of natural foods does not leak out unless some damage occurs to the tissues'. It is possible that the desiccation of fresh spinach in silica gel is too severe and membranes are lysed, allowing DNases access to compartmentalized DNA. On the other hand, air drying (see methods) may lead to more volatilization of water with less membrane damage. Even bound water (0.05 - 0.1 g H_2O per g dry wt) cannot be regarded as unavailable. Duckworth and Smith (1963) demonstrated that glucose, calcium chloride and sulfate migrated along food surfaces even at monolayer water values. These water activities (a_w = 0.2 - 0.3) were so low that, theoretically, water could not act as a solvent (Labuza, 1970). So it does appear that some degradative action may be occurring, even in desiccated materials. The problem of preserving specimens can be related to water activity (a_w). It is well established that the lower limits for growth of bacteria is a_W = 0.9 to 0.75 and for fungi 0.60, with no sustained growth of any known organism below a_W =0.60 (Bone, 1969). Labuza (1970) shows enzymatic activities approaching zero rates at a_W = 0.3, with non-enzymatic browning occurring down to a_W = 0.2 and lipid oxidation reaction rates minimized at a_W = 0.3, then **increasing at lower water activities!**. Unfortunately, research is needed to relate these principles to the preservation of DNA *in situ*.

The choice of desiccant was investigated and narrowed to drierite and silica gel. Although there are a number of compounds that are more powerful desiccants (such as P_2O_5), these compounds are so reactive with water to render them dangerous for routine field usage and particularly troublesome when trying to negotiate entry or exit with customs officials. Both

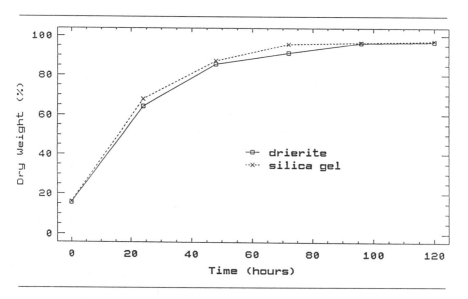

Figure 5. Changes in percent dry matter for fresh spinach leaves stored in drierite or silica gel. No significant differences were found between drierite and silica gel for any particular time period sample. See Table 1 for significance tests between time periods for both drierite and silica gel treatments.

drierite and silica gel are light, inert and inexpensive.
Tests of leaf drying (Figure 5) revealed that percentage of
dry matter asymptotes after 96 h at 96 to 97% dry matter (for
spinach) for both desiccants. Notice that the curves for both
drierite and silica gel are very similar. No significant differences in the percent dry matter were found between storing
samples in drierite or silica gel at any given time period.
No significant differences in the percent dry matter were
found after 72 h storage in drierite or silica gel (Table 1).

The dry matter values for spinach leaves compare with a
98.1% dry matter obtained by drying the moss, *Tortula ruralis*
with silica gel (Schonbeck and Bewley, 1981) and 96.4% dry
matter when *Selaginella lepidophylla* fronds were dried in

Table 1. Changes in dry weight for fresh spinach leaves when
stored in drierite or silica gel. All storage tests were performed at 22^0C. Any two means sharing a common superscript
are not significantly different by the SNK (Student-Newman-
Keuls) multiple range test, p = 0.05, based on 3 replicates.

Time stored	% dry weight after storage in	
	Drierite	Silica gel
Fresh (initial)	15.65[a]	16.05[a]
24 hours	64.28[b]	67.90[b]
48 hours	85.57[c]	87.27[c]
72 hours	91.23[c,d]	95.73[d]
96 hours	96.17[d]	96.52[d]
120 hours	96.77[d]	97.27[d]

drierite (Eickmeier, 1988). The relative humidity of drierite
is estimated at 1% and the water potential at -600 MPa (Eickmeier, 1988), compared to ca. 1% and -6000 MPa, respectively,
for silica gel (Schonbeck and Bewley, 1981). Wiebe (1981)
found *Tilia americana* leaves dried in a conventional plant
press reached a water activity (a_w) of 0.5 (-100 MPa), whereas
oven dried leaves (temperature and time not reported) had an
a_w of 0.25 - 0.35 (-150 to -200 MPa) and ground leaves, stored
over silica gel had an a_w of 0.06 (-400 to -500 MPa). Based

in Wiebe's graph of water potential and a_w (Wiebe, Figure 4, 1981), it appears that drierite is comparable to silica gel in desiccating power. Thus, it would seem that the levels of water activity attained by the use of drierite or silica gel (a_w less than 0.06) should stop both enzymatic and factors such as non-enzymatic browning but not stop lipid oxidation (Labuza, Figure 20, 1970). However, one should bear in mind that leaf materials are hydroscopic, so it is unlikely that the leaves reached the a_w of drierite or silica gel.

PROTOCOL FOR FIELD PRESERVATION OF FOLIAGE

Drierite has a water capacity of 10 to 14 percent, but above 6.6%, the capacity varies inversely with temperature (W. A. Hammond Drierite Co.). One would not want to risk possible rehydration of leaves, so storage ratios should be based on the 6.6% capacity. In lab tests, silica gel absorbed 8.85% of its weight of water after exposure to 100% humidity for 16 h at 22°C. We have found that plant materials contain as much as 92% moisture, so a useful approximation would be to assume the plant is mostly water and use 16 to 20 times the fresh leaf weight for the drierite or silica gel component.

Now that inexpensive ($100 USCY) battery powered, portable balances are available, one could take a supply of jars that hold (for example) 100 g of silica gel and then weigh out 5 g of fresh leaf material and add it to the jar along with silica gel (or drierite). We have found that air dried leaves (suitable for herbarium vouchers) generally contain from 10 to 15% water. Using a robust value of 20% water for air dried leaves, one can weigh out 5 g of air dried leaves (5 g x 20% = 1 g water) per 20 g of silica gel. This procedure may seem time consuming, but in practice, we merely do a quick check on the leaf area needed to give approximately 1 g (fresh leaves) or 5 g (dried leaves) and then just use that amount of leaf area. For example, for spinach, a 2 cm x 2 cm fresh leaf area weighs about 0.5 g. So, one can just cut the leaves into roughly 2 cm x 2 cm squares and add one square to 10 g of silica gel (0.5 g leaf / 10 g silica gel = 1/20 ratio). For succulent leaves, a slightly different protocol may be used. Liston *et al.* (1991) removed succulent leaf material after 24 hours in drierite and placed it in fresh drierite. A note of caution is necessary concerning field drying of specimens for subsequent silica gel/ drierite storage. We have experienced difficulty obtaining DNA from leaves dried at temperatures higher than about 55°C. In very rainy conditions where high drying temperatures (from butane stoves, for example) are used to dry specimens, it would seem advisable to merely blot leaves free of surface moisture and then place the fresh leaf material directly into silica gel or drierite. Liston et al., (1991) took 2-5 g of plant tissue and wrapped it in tissue paper to prevent it from fragmenting, then placed it in a 125 ml Nalgene bottle, 1/3 prefilled with drierite (with blue indicator crystals), and then filled the bottle (2/3) with additional drierite.

Plastic bottles are probably to be preferred to glass, to avoid breakage in transit. Using clear jars allows one to check the indicating crystals without opening the jar. The lids should be sealed with vinyl tape to insure against moisture leakage. The use of parafilm to seal containers is not recommended, as we have found it to come loose at 37°C (and of course, at tropical temperatures!).

Silica gel and drierite do differ in one characteristic that may be a consideration. We have found that silica gel can be dried (recharged) at 100°C for 24 h but drierite must be dried at much a higher temperature (we had to use 200°C). In addition, we could easily dry silica gel in a microwave oven, but were unable to dry (recharge) drierite in a microwave oven. If the desiccant gets wet before use, silica gel appears to be much easier to dry. Silica gel is used in large quantities for flower drying and, thus, may be cheaper, depending on your source. Neither drierite nor silica gel should be discarded but recharged for reuse on subsequent trips. If the materials are to be checked through customs, it is useful to have a small container of silica gel/ drierite that you can open and show the customs agents. A demonstration that the blue indicator crystals will turn pink when you breath on or moisten them is helpful in convincing the customs officials to **not** open your sealed specimen jars.

In conclusion, we have found that both fresh and air dried

leaves can be preserved in drierite or silica gel for up to 6
months at tropical temperatures (37°C) so plant materials can
be shipped by surface freight for the eventual extraction of
DNA. This should greatly facilitate the collection of materials since one will not have to use expensive and complex
freezing equipment in the field.

Acknowledgments - This research was supported by funds from
the Helen Jones Foundation, Baylor University Research Committee (grant 015-F87-UCR) and NSF (INT-8901632).

Literature Cited

Adams, R. P. 1988. The preservation of genomic DNA: DNA Bank-
Net. AIBS Meeting, Aug., Univ. of California, Davis, CA.
_____. 1990. The preservation of Chihuahuan plant genomes
through *in vitro* biotechnology: DNA Bank-Net, a genetic
insurance policy. Pp. 1-9. In: A. M. Powell, R. R. Hollander, J. C. Barlow, W. B. McGillivray and D. J. Schmidly
(eds.), *Third Symposium on Resources of the Chihuahuan
Desert Region*. Printech Press, Lubbock, TX.
Birren, D. W., E. Lai, L. Hood and M. I. Simon. 1989. Pulsed
field gel electrophoresis techniques for separating 1- to
50-kilobase DNA fragments. *Anal. Biochem*. 177: 282- 286.
Bone, D. P. 1969. Water activity - its chemistry and applications. *Food Prod. Devel.* Aug/Sept 1969, 81-94.
Bostock, C. J. 1988. Parameters of filed inversion gel electrophoresis for the analysis of pox virus genomes.
Nucleic Acids Res. 16: 4239- 4252.
Carle, G. F., M. Frank and M. V. Olson. 1986. Electrophoretic
separations of large DNA molecules by periodic inversion
of the electric field. *Science* 232: 65-68.
Chase, M. W. and H. H. Hills. 1991. Silica gel: An ideal mat-
erial for field preservation of leaf samples for DNA
studies. *Taxon* 40: 215-220.
Dally, A. M. and G. Second. 1989. Chloroplast DNA isolation
from higher plants: An improved non-aqueous method.
Plant Molec. Biol. Reptr. 7: 135-143.
Devos, K. M. and D. Vercruysse-Dewitte. 1989. Preparation of

plant DNA for separation by pulsed field gel electrophoresis. *Electrophoresis* 10: 267- 268.

Do, N. and R. P. Adams. 1991. A simple technique for removing
plant polysaccharide contaminants from plant DNA.
BioTechniques 10: 162-166.

Doyle, J. J. and J. L. Doyle. 1987. A rapid DNA isolation
procedure for small quantities of fresh leaf tissue.
Phytochemical Bull. 19: 11-15.

_____ and E. E. Dickson. 1987. Preservation of plant samples
for DNA restriction endonuclease analysis. *Taxon* 36:
715-722.

Duckworth, R. B. and G. Smith. 1963. The environment for chemical change in dried and frozen foods. *Proc. Nutr. Soc.*
22: 182.

Eickmeier, W. G. 1988. The effects of desiccation rate on
enzyme and protein-synthesis dynamics in the desiccation-
tolerant pteridophyte *Selaginella lepidophylla*. *Can. J.
Bot.* 66: 2574-2580.

Gekeler, V., S. Weger, E. Eichele and J. Probst. 1989. Computer controlled discontinuous rotating gel electrophoresis for separation of very large DNA molecules. *Anal.
Biochem.* 181: 227-233.

Labuza, T. P. 1970. Properties of water as related to the
keeping quality of foods. Proceedings, Third Congress
Food on Science and Technology. Pp. 618-635. Stewart and
Wiley, Wash. D.C.

Liston, A., L. H. Rieseberg, R. P. Adams, N. Do and G-L. Zhu.
1990. A method for the collecting dried plant specimens
for DNA and isozymes analyses, and the results of a field
test in Xinjiang, China. *Ann. Mo. Bot. Gard.* 77: 859-863.

Pyle, M. M. and R. P. Adams. 1989. *In situ* preservation of DNA
in plant specimens. *Taxon* 38: 576-581.

Higuchi, R., B. Bowman, M. Freiberger, O. A. Ryder and A. C.
Wilson. 1984. DNA sequences from the quagga, an extinct
member of the horse family. *Nature* 312: 282- 283.

Paabo, S. 1985. Molecular cloning of ancient Egyptian mummy
DNA. *Nature* 314: 644- 645.

_____. 1989. Ancient DNA: Extraction, characterization, molec-
ular cloning and enzymatic digestion. *PNAS.* (USA) 86:

1939-1943.

_____, J. A. Gifford and A. C. Wilson. 1988. Mitochondrial DNA sequences from a 7000-year old brain. *Nucleic Acids Res.* 16: 9775-9787.

Raven, P. H. 1987. Forests, people, and global sustainability. Keynote Address, National Audubon Society Biennial Convention, Western Washington University, Bellingham, WA.

_____. 1988. Tropical floristics tomorrow. *Taxon* 37: 549-560.

Rogers, S. O. and A. J. Bendich. 1985. Extraction of DNA from milligram amounts of fresh, herbarium and mummified plant tissues. *Plant Mol. Biol.* 5: 69-76.

Schonbeck, M. W. and J. D. Bewley. 1981. Responses of the moss *Tortula ruralis* to desiccation treatments. I. Effects of minimum water content and rates of dehydration and rehydration. *Can. J. Bot.* 59: 2698- 2706.

Shulman, S. 1986. Seeds of Controversy. *BioScience* 36: 647-651.

Wiebe, H. H. 1981. Measuring water potential (activity) from free water to oven dryness. *Plant Physiol.* 68: 1218-1221.

SOME POSSIBLE APPLICATIONS OF MOLECULAR GENETICS IN THE CONSERVATION OF WILD SPECIES FOR CROP IMPROVEMENT

T. Hodgkin and D. G. Debouck

IBPGR, c/o FAO
Via delle Sette Chiese 142
00145 Rome, ITALY

Summary - In recent years the possible use of wild relatives of crop species in plant breeding programs has been the subject of increased attention and this has stimulated the collection and *ex situ* maintenance of many wild plant species. These activities raise a number of questions concerning the number of wild species with which those interested in the conservation and utilization of plant genetic resources should be concerned, their relationships to crop species, the scale on which wild species' collection should be undertaken, the most suitable conservation methods for such material and the description of the genetic diversity available in these wider gene pools. Molecular genetic techniques have already provided useful information on such topics as the relationships between crop species and their wild relatives, the extent of genetic diversity in such species and genetic stability following *ex situ* storage. There remain significant opportunities for applying these techniques in studies of the distribution of genetic diversity in wild relatives of crop plants, and in the location, collection and maintenance of that diversity.

INTRODUCTION

The use of wild relatives of crop plants, and of other wild species, in crop improvement or the development of new crops has been the subject of a number of reviews over the past decade (Goodman *et al.*, 1987; Myers, 1983; Oldfield, 1989; Prescott-Allen and Prescott-Allen, 1983; Stalker, 1980) reflecting a wider recognition of the importance of wild relatives of crop species in plant breeding programs (e.g. Bonjean and Picard, 1990) and an increase in research in a number of key areas. One such area has been the evolution of crop plants and the relationship between different species within the same taxonomic group. Another has been the investigation of variation, especially that of agronomically important traits, in species which might be used in plant breeding programs. More general studies on the amount of variation in wild relatives of crop plants as compared with that in the crop have also been carried out (Doebley, 1989; Miller and Tanksley, 1990).

This interest in the use of wild species has been accompanied by an increased concern for their conservation. The attention given to safeguarding species and ecosystem diversity has been accompanied by work on the practical aspects of conserving and making available the genetic diversity within species. The conservation of wild relatives of crop species in such a way that they can be used for improving crop production raise a number of questions concerning the selection of material to be conserved, the development of appropriate conservation practices and improving the availability of these resources to plant breeders. While research on these questions has been in progress for some time, a number of important aspects have still to be resolved.

The rapid advances in molecular genetics and the development of a variety of techniques for the analysis of variation at the molecular genetic level were quickly recognized as having considerable potential in plant breeding programs. Work is now in progress in many laboratories throughout the world on the control and expression of potentially useful characters and on the development of gene transfer techniques to introduce such characters into crop cultivars (Weising et al., 1988). The potential use of molecular genetic techniques for conservation of wild relatives of crop plants has also been recognized. This is particularly true for research on the extent and distribution of variation within a species or between related species where molecular markers are considered to fulfill many of the criteria required for such studies. Thus molecular markers are easily detected, considered to be selectively neutral, can show high levels of polymorphism,

allow saturated coverage of the genome and their expression is
unaffected by the environment (Clegg, 1990). Other biochemical markers, such as seed proteins or isozymes, do not fulfill
these criteria to the same extent, although they can be easily
assayed in plant populations. Analysis of variation in plastid DNA also permits analysis of the maternal and paternal
contributions in evolutionary studies. As well as providing
new insights on evolution, species relationships and the variation present in different gene pools, molecular genetic techniques can assist conservation activities. In this paper we
intend to review some of the ways in which molecular genetic
methods have been applied to the investigation and conservation of the genetic diversity of wild relatives of crop plants
and to suggest some areas in which their use might be of most
benefit.

THE USE OF DIVERSITY AND THE NEED FOR DIVERSE COLLECTIONS

Four different ways in which wild species have been utilized can usefully be distinguished: the introduction or
transfer of species to new sites with potential for biomass
production; the development of new crops for the production of
novel products; the introduction into existing crops of useful
agronomic traits; and the development of improved breeding
methods.

The germplasm requirements of these approaches are quite
different and it may be useful to illustrate this with a few
examples. The introduction of new biotypes for biomass production has been frequently explored for forage production
systems. The possible use of *Stylosanthes* spp. in Australia
is an example of this approach (Skerman *et al.*, 1988). The
species are fast growing legumes from Africa and Latin America
which combine well with a range of grasses and are frequently
drought resistant. Where wild species are recognized to possess desirable products, breeding work has focused on the
identification and selection of attributes required for crop
production and for the desired product (Oldfield, 1989; Prescott-Allen and Prescott-Allen, 1983). Examples of wild
species which are being or have recently been domesticated
include jojoba (*Simmondsia chinensis*) which provides an eco-

nomic substitute for sperm whale oil and which is now grown
commercially on an extensive scale and a number of other oil
producing species such as *Crambe abbysinica* and species of
Limnanthes (Oldfield, 1989). In addition to oil-producing
species, those producing pharmaceutically active compounds are
also the subject of considerable interest. A well known
example is the Madagascar Perwinkle (*Catharanthus roseus*)
which contains vincristine and vinblastine, two compounds with
potential in the treatment of cancer.

The most common and best known use of wild relatives is in
the introduction of desirable agronomic trails, most commonly
disease resistance or stress tolerance. Good examples of this
are provided by the tomato and potato. The relative genetic
uniformity of these two crops as cultigens (for tomato see
Rick and Fobes, 1975, Rick, 1976; for potato see Hawkes,
1978), the possibilities for wide crossing (for tomato: Rick,
1978, 1988; for potato: Ross, 1978), their economic importance
(Harlan, 1984) and the early existence of relatively diverse
collections (Esquinas-Alcazar, 1981; Hawkes, 1990) probably
account for this extensive use.

The introduction of desirable traits from wild relatives
depends primarily on the ease with which the wild taxa and
crop species can be intercrossed. A wide variety of different
possible relationships exist between crop plants and their
wild relatives depending on the nature of the species concerned and the origin and evolution of the crop. A flexible
and practical approach, of direct use to plant breeders, to
the classification of the different kinds of relationships
that can exist was developed by Harlan and de Wet (1971).
They distinguished three kinds of relationships between wild
relatives and crop species which they identified as belonging
to primary, secondary or tertiary gene pools. The primary
gene pool (GP1) includes the biological species, i.e. the crop
and all relatives with which the crop is fully cross compatible. Thus for barley, both *Hordeum vulgare* and *H. spontaneum* belong to GP1 while other *Hordeum* species, except possibly *H. bulbosum*, belong to the tertiary gene pool. Secondary
and tertiary gene pools (GP2 and GP3) include species of
increasing incompatibility. Crosses between GP2 species and

the crop are largely infertile setting few seeds while those
involving GP3 fail to produce any seed and can usually only be
crossed successfully by artificial techniques involving tissue
culture or by protoplast fusion.

Although plant breeders would prefer to find useful traits
in the primary gene pool, they may need to extend their search
to the secondary gene pool, and in some cases, even to the
tertiary gene pool to find the desired characters. Several of
the examples noted below involve species considered to belong
to secondary gene pools. In tomatoes, L. esculentum var. cer-
asiforme is the closest relative of the cultivated tomato
(GP1) and has been extensively used in crossing programs, all
other Lycopersicon species belong to GP2 or GP3 and have also
been used (Warnock, 1988). In the case of potato, the value
of other series of Solanum subgenus Potatoe such as Demissa,
Yungasensa, Tuberosa, as detailed by Hawkes (1990), has been
early recognized for potato breeding. It is worth noting that
even when previous attempts to cross two species have failed,
future work may enable such crosses to be made and the useful
genes in the wild relative to be utilized. Rick (1988)
reported that successful crosses had been made between L.
esculentum and Solanum lycopersicoides despite the failure of
earlier attempts to obtain fertile hybrids. Iwanaga et al.
(1991) have recently succeeded in crossing S. tuberosum with
S. acaule.

Tolerance to environmental stress has often been obtained
from germplasm submitted to stress over long periods of time.
Thus, tolerance to salinity was found in L. cheesmanii, a wild
tomato distributed on the sea shores of the Galapagos Islands
(Rick et al., 1987) and frost tolerance for potato was identified in S. acaule distributed around 3,500-4,200 m in the
highlands of Bolivia (Hawkes and Hjerting, 1989). In a similar way, it seems likely that coevolution between pathogens
and the host plant increases the probability of finding pathogen resistance in wild relatives. The wild tomato, L. hirsutum, from moist Andean Ecuador and Peru, possesses resistance
to several fungal diseases which are endemic to the area
(Rick, 1988). Sources of resistance to potato late blight
were identified in S. demissum, a wild potato from the Eje

Volcanico, Mexico [where the sexual cycle of the fungus takes place (Rowe, 1969)] and Cuchumatanes, Guatemala (Hawkes, 1966). One should note that the cultigens, which are susceptible to the fungus, are of Andean origin (Grun, 1990). Similarly, *S. stoloniferum* distributed in central Mexico (Hawkes, 1966) was found to be an excellent source of resistance to potato viruses A and Y (Ross, 1978). However, there is no guarantee that this will necessarily always be the case and reviewing the evidence on the occurrence of resistance genes in plant species Harlan (1978) concluded that "resistance is where you find it".

Even where a particular place or population is known as a source of a useful trait, not all plants in that population may possess that trait. Certain variants of arcelin, a lectin-like seed storage protein conferring resistance to weevils (Osborn et al., 1988), were found in only some plants of some populations of the whole range of *Phaseolus vulgaris* (Osborn et al., 1986). Resistance to the grassy stunt virus in rice was found in only one accession of *Oryza nivara* (Khush and Ling, 1974).

While resistance to diseases, pests and adverse environmental factors are the characters most commonly sought in wild germplasm, other desirable traits have also been detected and used. *Triticum turgidum* var. *dicoccoides*, a wild relative of wheat from the eastern Mediterranean, was used to increase the grain protein content of bread wheat and durum wheat without losing high yield (Levy and Feldman, 1987). In some cases the results of using wild relatives can be unexpected : *Gossypium thurberi*, a wild cotton from NW Mexico, has no lint but was used to increase lint strength in upland cotton (Harlan, 1976).

Wild relatives of crop plants have also been used to develop new breeding methods. Cytoplasmic male sterility has been induced by introducing an alien cytoplasm into a species. Examples include combining *Gossypium anomalun* or *G. arboretum* cytoplasm with the *G. hirsutum* genome (Simmonds, 1979) and the use of a *Raphanus raphanistrum* 'S.' cytoplasm to produce male sterility in *Brassica oleracea* and *B. napus* (Bannerot et al., 1974). In most cases such procedures have not yet been used

to produce commercial cultivars owing to practial problems
resulting from genome/cytoplasm incompatibilities.

Another use of wild species has been in the production of
homozygous lines in barley.   *Hordeum bulbosum* can be used in
crosses with *H. vulgare* to produce haploid *H. vulgare* plants
which, following chromosome doubling with colchicine, result
in homozygous diploids (Bonjean & Picard, 1990).   In practice,
when possible, anther culture may now provide a more effective
method of producing such homozygotes.

One important conclusion that can be drawn from this brief
summary is that there are no fixed rules to predict where useful germplasm, traits for further breeding or novel breeding
techniques will be found or indeed what will prove to be useful in the future.   The safest longterm insurance for those
involved in the conservation and utilization of plant genetic
resources would seem to be to try and conserve maximum diversity of those wild relatives most likely to be of interest in
crop improvement.   Such a strategy requires:

- An understanding of species relationships and of the
  evolution of crop plants (phylogenetic relationships).
- A knowledge of the amount of diversity present in different gene pools in relation to particular crop
  species.
- Information on the distribution of diversity within
  gene pools with respect to climatic, ecological and
  geographical factors.
- The development of techniques that will improve the
  effectiveness of both *in situ* and *ex situ* conservation
  strategies.

Molecular genetic studies have already begun to make significant contributions in these four areas.   While it would be
premature to undertake a full scale examination of the contribution that molecular genetics can make in the conservation of
the plant genetic resources of wild relatives of crop plants,
it is useful to outline the contribution that such research
has already made to the different areas listed above.   This
will permit identification of some of the major topics where
such techniques may prove particularly useful in the future.

PHYLOGENETIC RELATIONSHIPS

As noted above, the primary gene pool (GP1) includes the biological species, i.e. the crop and all relatives with which there is full genetic compatibility, while GP2 and GP3 include materials of increasing incompatibility. Biochemical markers such as seed storage proteins have been useful in clarifying relationships within GP1 for beans, lentils, chickpea, pigeon-pea, watermelon and peanut (Gepts, 1990), while isozyme studies have been used for banana, barley, foxtail millet, maize, quinoa, rice and tomato (Doebley, 1989). Molecular markers have been used to clarify these relationships further to extend research to include GP2 and GP3 taxa, particularly by examining variation in plastid DNA. In the Andean potato, *Solanum tuberosum* ssp. *andigena*, similarities in cpDNA sequences (Hosaka and Hanneman, 1988) have confirmed the early assumption that forms of *S. stenotonum* constitute one of the ancestors of the cultivated species, but as discussed by Grun (1990), the other parent is still disputed. For the Andean gene pool of the common bean - a subunit of the GP1 since partial incompatibility has been found within *Phaseolus vulgaris* (Gepts and Bliss, 1985; Sprecher, 1988) - RFLP analysis of mtDNA points to specific places of domestication in southern Peru and Bolivia (Khairallah et al., 1991), where studies with protein markers were inconclusive (Vargas et al., 1990). The analysis of cpDNA data of the genus *Zea* (Doebley, 1990) not only confirms that subsp. *parviglumis*, a wild teosinte from the Balsas region in southwestern Mexico, is likely to be involved in the ancestry of Indian corn confirming earlier isoenzyme evidence (Doebley et al., 1984), but also that some of the other teosintes, *Z. perennis*, *Z. diploperennis* and *Z. luxurians*, were not involved in the ancestry but evolved independently. However, the analysis failed to separate the annual teosintes of subsp. *mexicana* from those of subsp. *parviglumis*, although they had previously been shown to differ on the basis of 12 enzyme systems (Doebley et al., 1984). This indicates a possible limitation in the use of cpDNA data which has often shown rather limited variation. An analysis of cpDNA variation in *Pisum sativum*, *P. humile*, *P. elatius* and *P. fulvus* showed that *P. fulvus* was the most distant taxon, and

that several pea cultivars were derived from *P. humile*, traditionally considered as the ancestral form (Zohary and Hopf,
1973). However, the possibility that *P. elatius* was involved
at some stage in the evolution of the crop cannot be ruled out
on the cpDNA data (Palmer *et al.*, 1985). A study of cpDNA
variation in *Oryza* spp. (Dally and Second, 1990) confirmed
that African rice, *Oryza glaberrima*, was domesticated from *O.
breviligulata*, an annual endemic African wild rice (Second,
1985), independently from Asiatic rice, *O. sativa*. The cpDNA
study (Dally and Second, 1990) confirmed that the two *O.
sativa* groups, *indica* and *japonica* previously distinguished by
isozyme data (Second, 1982), could be traced to two groups of
wild *O. rufipogon*, one distributed in South Asia and another
one in China respectively (Second, 1985). The data also indicated that several divergences in the *O. rufipogon* complex had
occurred prior to domestication.

RFLP analysis on nuclear DNA in *Lycopersicon* species (Miller and Tanksley, 1990) has helped confirm the close relationships between *L. esculentum*, *L. pimpinellifolium* (both very
closely related probably due to introgression where they are
sympatric) and *L. cheesmanii*, apart from the other group of
green-fruited species. RFLP analysis has also revealed striking similarities in genome organization between *L. esculentum*
and *S. tuberosum*, demonstrating that for 9 of the 12 chromosomes in the two species the order of the identified loci was
identical (Bonierbale *et al.*, 1988). It has also been used to
explore relationships in such complex crop plant genera as
Brassica (Song *et al.*, 1988; 1990) and *Triticum* (Talbert *et
al.*, 1991).

It seems likely that in this area of research the study of
RFLPs in nuclear or plastid genomes and of repetitive DNA such
as ribosomal DNA genes (Dvorak, 1990) will rapidly extend the
amount of information available, providing insights not only
on the extend of relationship between species but also on the
kinds of genetic rearrangements that have occurred. As
studies of cpDNA variation in *Oryza* spp. have shown, such
research can also provide new insights into the evolution and
possible time of divergence of different species (Dally and
Second, 1990). In fact, molecular genetic data is uniquely

suitable for such studies and further information will come
from detailed analysis of variation in particular gene
sequences in related species.

THE AMOUNT OF GENETIC DIVERSITY PRESENT

As well as obtaining some estimate of the amount of variation present in a species, it is important to develop effective strategies for the use of wild relatives, to obtain
information on the relative amounts of variation present in
crop species and their wild relatives. Data on the extent to
which variation at specific loci is absent in the crop but
present in wild relatives is also desirable. An evaluation of
the true genetic basis of a crop species and their wild relatives is of direct use in developing sustainable, less vulnerable crop production (Jackson and Ford-Lloyd, 1990; National
Research Council, 1972). Furthermore, since the number of
accessions in a germplasm bank cannot be indefinitely
increased, germplasm banks may wish to concentrate resources
on accessions that represent a more complete fraction of
genetic diversity.

Evidence from a number of sources suggest that there may
be considerable differences in the amount of genetic variation
present in different species. While differences in techniques
make direct comparison difficult (Graner et al., 1990) and the
number of accessions tested in any species remains small, it
seems that certain species contain much more RFLP polymorphism
than others (Shattuck-Eidens et al., 1990).

Clearly, molecular markers are the markers of choice for
such studies which should be selection neutral and environment
neutral. Morphological markers tend to overestimate diversity
in the cultigens because of man oriented selective pressures
keeping certain recessive alleles at relatively high frequencies. Striking examples of this are the Indian amaranths
where morphological variation contrasts with almost monomor-
phic allozyme loci (Jain et al., 1980), and *Gossypium barba-
dense* where the extensive variation in fibre colours contrasts
with low isoenzyme variation (Bourdon, 1986). Biochemical
markers (such as seed storage proteins, isozymes, RFLPs) seem
to generally satisfy these conditions. Although, as stressed

by Stebbins (1990), the neutrality and random characteristics of the markers involved should be carefully checked, it seems that environmental factors do not influence electrophoretic patterns of seed proteins (Gepts, 1990). Thus, the variants of arcelin in some populations of wild bean appear to be selection neutral inasmuch they are restricted to few individuals in a small part of the wide range of distribution (Toro et al., 1990) and are absent from most populations of wild beans.

Studies to compare the amounts of diversity present in crop species and their wild relatives or ancestors suggest that two situations may occur. In some crops it would appear that much more diversity occurs in the crop than in its immediate ancestor while in most species it seems that the ancestor or close relatives contain extensive variation not present in the crop.

Examples of the former situation are found with maize and possibly other outbred crop species such as Brassica oleracea. The case of maize is somewhat extreme. While there is little doubt that there is a considerable variation in the crop at both morphological (e.g. Wellhausen et al., 1952) and biochemical levels (e.g. Doebley et al., 1985; Goodman and Stuber, 1983; Helentjaris et al., 1985), a strict comparison with its ancestor is difficult, as a formal ancestor has still to be defined even if the teosinte from the Balsas region has probably involved at some stage (Doebley, 1990).

Examples of the latter situation are found in Phaseolus beans, where with the help of biochemical markers (variants of seed storage proteins revealed by SDS-PAGE) more diversity has been found in the wild ancestral forms (for common bean: Gepts et al., 1986; for lima bean: Debouck et al., 1989; for tepary bean: Schinkel and Gepts, 1988; for P. polyanthus bean: Debouck, et al., 1990). A current explanation to this "founder effect" is in the domestication process, a few wild plants being actually domesticated (Ladizinsky, 1985). Some considerable variation is apparent between bean species (Table 1), and between regions and hence between groups of cultivars within species, modifying substantially the importance of this effect (Debouck and Tohme, 1989; Maquet et al., 1990).

Table 1. Quantitative estimation of the founder effect in *Phaseolus* beans by the number of variants in seed proteins (sources: see text).

Species	Region	Wild forms	Cultivated forms
P. vulgaris	Mesoamerica	16	2
P. vulgaris	Andes	13	7
P. lunatus	Mesoamerica	7	2
P. lunatus	Andes	5	4
P. acutifolius	Mesoamerica	14	1
P. polyanthus	Mesoamerica	6	4

In the case of the chili pepper *Capsicum annuum*, although the wild races var. *aviculare* are widespread from Arizona down to northern Peru (McLeod *et al.*, 1983), it seems on the basis of karyotype evidence (Pickersgill, 1971) and multivariate analysis (Pickersgill, 1984) that domestication would have taken place somewhere in central Mexico and from a "rather small sample of wild plants". However, after an electrophoretic study, McLeod *et al.* (1983) still consider the possibility of multiple, independent domestications of local forms from a more variable wild gene pool. Molecular genetic studies may provide the evidence necessary to resolve this question.

In *Lycopersicon* molecular genetic studies of single copy nuclear RFLPs have shown that both within accession and between accession variation is greatest in the wild self incompatible species *L. hirsutum*, *L. penellii* and *L. peruvianum* (Miller and Tanksley, 1990). Almost 75% of the unique restriction fragments found in the study came from these three species while *L. esculentum* possessed only 5% of the unique fragments. Studies in rice of rDNA spaces length polymorphism have also shown that the wild species *O. rufipogon* and *O. longistaminata* contain many more size classes than do cultivated *O. sativa* and *O. glaberrima* (Cordesse *et al.*, 1990).

These studies, like those on pearl millet DNA sequence variation (Gepts and Clegg, 1989) or on RFLP variation in soyabean
and its wild perennial relatives (Menancio *et al.*, 1990) dem-
onstrate the importance of wild relatives of crop species.
However, to date research has, at the molecular genetic level,
been based on rather few accessions and the genetic basis full
significance of the variation detected is largely unknown.
Further studies using a wide range of carefully chosen accessions are needed which should also examine variation of gene
loci known to show differing rates of evolution.

THE SPATIAL DISTRIBUTION OF DIVERSITY WITHIN GENE POOLS

Genetic diversity is not distributed at random in wild
populations, and its organization depends, among other factors, on the taxonomic status of the species involved, its
life form, breeding system, seed dispersal mechanisms, and
successional status (Hamrick *et al.*, 1979). Current evidence
suggests that geographic range accounts for most of the
observed variation (Hamrick and Godt, 1990) and in many of the
species surveyed, clear clinal gradients have been found. For
Hordeum spontaneum, half of the alleles surveyed in studies of
allozyme variation were not widespread but concentrated in one
particular region of the Near East Fertile Crescent (Nevo *et
al.*, 1986a). This study also showed that some allozyme
frequencies were significantly associated with combinations of
temperature, rainfall and evaporation (Nevo *et al.*, 1979),
soil types (Nevo *et al.*, 1981) and exposure (Nevo *et al.*,
1986b). Similar results were obtained on *Avena barbata*, a
weed introduced from Spain both in California (Clegg and
Allard, 1972; Hamrick and Allard, 1972; Hamrick and Holden,
1979) and in Israel (Kahler *et al.*, 1980). In wild common
beans, the populations of which cover a long distribution arc
in the Americas (Toro *et al.*, 1990), important regional variation in seed proteins (phaseolin: Gepts *et al.*, 1986; Gepts
and Bliss, 1986; Vargas *et al.*, 1990; arcelin: Osborn *et al.*,
1988) and in allozymes (Koenig and Gepts, 1989) has been
described, with variants being found specifically in certain
parts of that range.

Information from molecular genetic studies on the distri-

bution of diversity and the effect that geographic climatic or
ecological factors may have on that variation is currently
extremely limited. Partly this reflects the labor involved in
testing the large number of samples that would be involved in
such studies. Some inferences may be drawn from other
research and investigations on ribosomal DNA polymorphism in
Hordeum vulgare and *H. spontaneum* have shown that certain
variants are favored by Mediterranean climates while other
variants are strongly selected against (Saghai-Maroof *et al.*,
1984; Zhang *et al.*, 1990). Whether this involves direct
selection of the loci involved or some kind of "hitch-hiking"
effect is unknown. Until methods for processing large numbers
of samples quickly and cheaply become available, it is
unlikely that extensive molecular genetic studies will be
undertaken in this area.

TECHNIQUES FOR CONSERVATION

As well as contributing to our knowledge of relationships
between species, their diversity and its distribution, molecular genetics is beginning to make a direct contribution to the
activities involved in conserving wild relatives of crop
plants. However, there are still relatively few published
reports in this area and many of the possible applications of
molecular genetic techniques have still to be tested.

While *ex situ* conservation has been used for many wild
relatives of crops, and extensive collections have been built
up for some species, it is likely that both *in situ* and *ex
situ* conservation will be used to safeguard these genetic
resources. Indeed for many wild relatives of crops, *in situ*
conservation may well be the most appropriate strategy. The
numbers of species of wild relatives of crop plants is so
great (possibly as many as 2,500) that *ex situ* conservation of
all of them would be impossible, and in case of long lived
perennial species or those with recalcitrant seeds, *in situ*
conservation may be the only practical option.

The identification of sites and populations for *in situ*
conservation will involve surveys of the target species to
identify areas of rich diversity, determine effective population sizes and species breeding system. To date, isozyme

variation has been commonly used for studies of natural diversity in wild relatives (e.g. *Hordeum spontaneum*, Jana and
Pietrzak, 1988) and for work on the extent of cross and self-
pollination (Brown *et al.*, 1989). There is little or no
information from molecular genetic studies on diversity in
natural populations although some deductions can be made from
RFLP studies or the investigation rDNA variation such as those
undertaken by Miller and Tanksley (1990) for *Lycopersicon* or
by Zhang *et al.* (1990) for *Hordeum spontaneum*. Such studies
should include research on the changes that may occur in gene
frequency with time in target populations. Data on this
aspect are extremely limited except for artificial populations. Saghai-Maroof *et al.* (1984) investigated changes in
the frequency of rDNA genes in a cultivated barley population
derived from a diverse set of lines over 53 generations and
found significant changes in the identity and frequency of the
commonest alleles and loss of five alleles, present in the
original population, by the F_{13} generation.

Ex situ conservation involves collection of desired accessions, their storage in such a way as to maintain their
genetic integrity and their regeneration when appropriate. It
will normally also involve multiplication for distribution.
In order to carry out these processes successfully, the accessions should be maintained free of disease and the health status of *ex situ* samples is therefore also important.

Collecting wild species for *ex situ* storage presents for-
midable difficulties and there has been considerable discussion of the best procedures to follow (Marshall, 1990). Practical difficulties include the scattered distribution of wild
species in comparison with crops, the great variation that may
exist in flowering and fruiting times and the small numbers of
seed that can be collected. In some cases seed collection may
not be possible and *in vitro* tissue collecting techniques have
been developed. These are equally applicable to wild species
although, as with seed collection, major difficulties can
occur due to the wide distribution of the target material and
to its lack of accessibility.

One approach to improving the efficiency of collecting
involves the use of preliminary ecogeographical surveys to

guide collecting activities (Anonymous, 1985). Such surveys
vary considerably in scope and may or may not be linked with
collecting activities (Mukherjee, 1985; Nabhan, 1990). There
is potential to extend the value of such work by investigating
variation in plant tissue collected from chosen specimens and
analyzed for isozyme or RFLP variation (Pyle & Adams, 1989).
An important component of any ecogeographic survey is the
study of herbarium material of the target species prior to
collection of samples for *ex situ* conservation. Techniques
have been developed for extraction of DNA from herbarium specimens (Rogers & Bendich, 1985) and further studies on the use
of these in the context of ecogeographic surveys would clearly
be valuable.

Ex situ storage of seeds or plant tissue *in vitro* requires
that the stored sample is maintained as long as possible with
a minimum loss of viability and without any change in its
genetic constitution. Accessions of wild species are no different from crop samples in this regard except that seed
sample sizes are often small and little may be known about
optimum storage conditions. It has long been known that prolonged storage of seed can cause genetic damage but the extent
and effect of this damage appears to be variable. Recent
studies using molecular probes to investigate the extent and
nature of damage at the DNA level have shown that damage can
be detected using RFLPs and other techniques (Osborne, 1982;
Elder *et al.*, 1987). Where genetic damage results from sub-
optimum storage conditions, it may occur more frequently in
wild germplasm accessions.

In vitro conservation methods have not been extensively
applied to wild relatives of crop plants because in many cases
wild species are fertile, producing seed, even when their crop
relatives are sterile or almost so as in the case of yam and
sweet potato. *In vitro* maintenance may also be used in some
crops which are normally clonally propagated in order to maintain a particular genotype even when seed is available (e.g.
Rubus, potato), a consideration that is rarely of importance
in the conservation of wild species. Where *in vitro* techniques are appropriate, as in long lived perennials and
species with recalcitrant seeds or sterility systems, there is

a need to determine genetic stability following storage and, as in cultivated species, molecular genetic procedures such as RFLP or fingerprinting techniques using repetitive DNA sequences could be valuable. Such techniques have been applied to assess stability following tissue culture in crop plants.

Regeneration of stored seed samples is recommended whenever the viability of the sample drops below a critical level. The objective of regeneration is to obtain a fresh supply of seed while maintaining the genetic integrity and diversity of the accession. There are major problems in carrying out successful regeneration of any sample. Loss of variation may occur due to "bottlenecking" and shifts in gene frequency can result from the fact that the environment in which regeneration takes place is frequently very different from that in which the sample originated (Breese, 1989). Regeneration of wild species presents additional problems. Seed production may be very low, natural seed dispersal mechanisms make seed harvesting difficult and seed maturation may occur over several weeks. In addition, regeneration of wild species takes place outside the natural environment of the species and this may well have a significant effect on gene frequencies. Few detailed studies have been carried out, even for many crops species, and it has been recognized that this is an area where research is urgently required (Breese, 1989).

Molecular genetic techniques have yet to be used in studies on the integrity of conserved samples and it is likely that their use will be limited to situations in which earlier studies of diversity or allele distribution have shown that they provide useful information. In self-pollinating species, where accessions are essentially homogenous, some techniques such as DNA fingerprinting may also be useful to confirm the purity of the regenerated accession avoid maintaining unnecessary duplicates and determine the precise parentage of accessions (e.g. Nybom, 1990).

Maintaining healthy stocks of germplasm is an essential part of *ex situ* conservation both to ensure safe maintenance of accessions and to satisfy the quarantine requirements that will be encountered when germplasm is distributed. In this latter respect, the increased use or wild relatives in crop breeding causes health problems similar to those involved in using any exotic germplasm with the added dimensions that information on known or potential pathogens may be limited or absent. Since alien germplasm may introduce susceptibility to new pests or diseases there is good reason to be concerned.

One approach to this problem would be the development of genetic tests for whole classes of pathogens or potential pathogens. Tests which detected a wide range of potential pathogens, even if their identity and precise characteristics were unknown, would clearly be of great benefit. It may well be possible to develop DNA probes which would recognize coat protein genes or other genes common to a number of different viruses. Such an approach may also be possible for bacterial pathogens but it is unlikely that probes of such a general nature could be developed for fungal pathogens.

CONCLUDING REMARKS

Molecular genetic techniques have already provided valuable data relevant to the conservation of wild relatives of crop species. The techniques currently in use have proved especially useful for the analysis of phylogenetic relationships and in charting the evolutionary process more fully. They are also producing new information on the amount of genetic variation present in species of interest. Little information has yet to be published on the distribution of diversity at the molecular genetic level. The techniques have also been used in some areas of conservation work and seem likely to prove particularly useful in studies on the integrity of stored germplasm and for the location of diversity so that both *in situ* and *ex situ* conservation can be targeted more effectively to those species and population with greatest diversity.

It would be unwise to attempt to predict in what areas molecular genetic techniques are likely to prove most useful in the future. New technologies are being rapidly developed so that new genome components such as the copia-like elements found in *Pisum* (Lee *et al.*, 1990) become available for analysis of diversity and new techniques for screening diversity can be applied such as random amplified polymorphic DNA (RAPDs) using PCR technologies (Anderson and Fairbanks, 1990). Nonetheless, it is likely that DNA technologies will only achieve their full potential in conservation genetics when techniques are developed whereby large samples (500-1,000) can be screened without radioactive chemicals at relatively low costs. At present the limitations of molecular genetic data are not usually related to the techniques themselves but rather to the quality and extent of the samples used.

The investigation of the distribution of diversity in populations at the DNA level is an important area for further research. There is also a need to explore the practical use of DNA technologies in identifying duplicates in germplasm banks and locating new genetic diversity for conservation. As new technologies are developed new areas for research will undoubtedly become apparent, particularly in respect of the use of variation in particular DNA sequences to study crop plant evolution and the maintenance of diversity.

As discussed above, there is already abundant evidence of the need to preserve a wide range of diversity to meet current crop improvement needs. There is also clear evidence that conservation of the widest range of genetic diversity is important to meet future, as yet unidentified, needs. Thus, Doney and Whitney (1990) report that *Beta maritima* genetic resources collected in Europe in 1925 and 1935 have been shown to contain useful resistance to Rhizomania caused by BNYV virus, *Erwinia*, sugar beet root maggot and *Cercospora* which are or have recently become important diseases in USA. This striking demonstration that such resources can become useful underlines the need to preserve the widest possible diversity of wild relatives of crop plants. Molecular genetics provides a range of powerful techniques which have considerable potential to help those concerned to conserve genetic diversity to identify resources likely to contain the widest possible diversity. The techniques can also help in making that conservation more effective and more useful for crop production.

Literature Cited

Andersen, W. R. and D. J. Fairbanks. 1990. Molecular markers:
    important tools for plant genetic resource characteriza-
    tion. *Diversity* 6: 51-53.
Anonymous 1985. Ecogeographical surveying and *in situ* conser-
    vation of crop relatives. International Board for Plant
    Genetic Resources, Rome, Italy. 27p.
Bannerot, H., L. Loulidard, Y. Cauderon, and J. Tempe. 1974.
    Cytoplasmic male sterility transfer from *Raphanus* to *Bras-
    sica*. Pp. 52-53. *In*: A. B. Wills and C. North (eds.),
    Eucarpia meeting-Cruciferae 1974. Scottish Horticultural
    Research Institute. Dundee, United Kingdom.
Bonierbale, M. W., R. L. Plaisted and S. D. Tanksley. 1988.
    RFLP maps based on a common set of clones reveal modes of
    chromosomal evolution in potato and tomato. *Genetics* 120:
    1095-1103.
Bonjean, A. and E. Picard. 1990. *Les cereales a paille Orig-
    ine, histoire, conomie, slection*. Softword/ITM and
    Groupe Verneuil, Verneuil l'Etang, France. 205p.
Bourdon, C. 1986. Polymorphisme enzymatique et organisation
    gntique des deux espces cultives tetraplodes de
    cotonnier, *G. hirsutum* et *G. barbadense*. *Cot. Fib. Trop.*
    41: 191-210.
Breese, E. L. 1989. *Regeneration and multiplication of germ-
    plasm resources in seed genebanks: the scientific back-
    ground*. International Board for Plant Genetic Resources,
    Rome, Italy. 69p.
Brown, A. H. D., J. J. Burdon and A. M. Jarosz. 1989. Isozyme
    analysis of plant mating systems. Pp. 73-86. *In*: D. E.
    Soltis and P. S. Soltis (eds.), *Isozymes in plant biology*.
    Dioscorides Press, Portland, Oregon, USA.
Clegg, M. T. 1990. Molecular diversity in plant populations.
    Pp. 98-115. *In*: A. H. D. Brown, M. T. Clegg, A. L. Kahler
    and B. S. Weir (eds.), *Plant population genetics, breed-
    ing, and genetic resources*. Sinauer Associates Inc., Sun-
    derland, Massachusetts, USA.
_____ and Allard, R. W. 1972. Patterns of genetic differentia-
    tion in the slender wild oat species *Avena barbata*. *Proc.*

Natl. Acad. Sci. USA. 69: 1820-1824.

Cordesse, F., G. Second and M. Delseny. 1990. Ribosomal gene spacer length variability in cultivated and wild rice species. *Theor. Appl. Genet.* 79: 81-88.

Dally, A. M. and G. Second. 1990. Chloroplast DNA diversity in wild and cultivated species of rice (Genus *Oryza*, section Oryza). Cladistic-mutation and genetic-distance analysis. *Theor. Appl. Genet.* 80: 209-222.

Debouck, D. G., A. Maquet and C. E. Posso. 1989. Biochemical evidence for two different gene pools in lima beans, *Phaseolus lunatus* L. *Annu. Rept. Bean Improvement Coop.* 32: 58-59.

_____, V. Schmit, D. Libreros Ferla, and H. Ramirez. 1990. Biochemical evidence for a fifth cultigen within the genus *Phaseolus*. *Annu. Rept. Bean Improvement Coop.* 33: 106-107.

_____ and J. Tohme. 1989. Implications for bean breeders of studies on the origins of common beans, *Phaseolus vulgaris* L. Pp. 3-42. In: S. Beebe (ed.), *Current topics in breeding of common bean*. Bean Program, Centro Internacional de Agricultura Tropical, Cali, Colombia.

Doebley, J. 1989. Isozymic evidence and the evolution of crop plants. Pp. 165-191. In: D. E. Soltis and P. S. Soltis (eds.), *Isozymes in plant biology*. Dioscorides Press, Portland, Oregon, USA.

_____. 1990. Molecular evidence and the evolution of maize. *Econ. Bot.* 44(3 Supplement): 6-27.

_____, M. M. Goodman and C. W. Stuber. 1984. Isoenzymatic variation in *Zea* (Gramineae). *Syst. Bot.* 9: 203-218.

_____, _____ and _____. 1985. Isozyme variation in the races of maize from Mexico. *Amer. J. Bot.* 72: 629-639.

Doney, D. L. and E. D. Whitney. 1990. Genetic enhancement in Beta for disease resistance using wild relatives: a strong case for the value of genetic conservation. *Econ. Bot.* 44: 445-451.

Dvorak, J. 1990. Evolution of multigene families: the ribosomal RNA loci of wheat and related species. Pp. 83-97. In: A. H. D. Brown, M. T. Clegg, A. L. Kahler and B. S. Weir (eds.), *Plant population genetics, breeding and genetic resources*. Sinauer Associates, Inc., Sunderland, Massachusetts, USA.

Elder, R. H., A. Dell'Aquila, M. Mezzina, A. Sarasin and D. J. Osborne. 1987. DNA ligase in repair and replication in the embryos of rye, *Secale cereale*. *Mutation Res.* 181: 61-71.

Esquinas-Alcazar, J. T. 1981. *Genetic resources of tomatoes and wild relatives - a global report*. International Board for Plant Genetic Resources, Rome, Italy. 65p.

Gepts, P. 1990 . Genetic diversity of seed storage proteins in plants. Pp. 64-82. In: A. H. D. Brown, M. T. Clegg, A. L. Kahler and B. S. Weir (eds.), *Plant population genetics, breeding and genetic resources*. Sinauer Associates Inc., Sunderland, Massachusetts, USA,

_____ and F. A. Bliss. 1985. F_1 hybrid weakness in the common bean - Differential geographic origin suggests two gene pools in cultivated bean germplasm. *J. Hered*. 76: 447-450.

_____ and _____. 1986. Phaseolin variability among wild and cultivated common beans (*Phaseolus vulgaris*) from Colombia. *Econ. Bot*. 40: 469-478.

_____ and M. T. Clegg. 1989. Genetic diversity in pearl millet (*Pennisetum glaucum* (L.) R. Br.) at the DNA sequence level. *J. Hered*. 80: 203-208.

_____, T. C. Osborn, C., K. Rashka and F. A. Bliss. 1986. Phaseolin protein variability in wild forms and landraces of the common bean (*Phaseolus vulgaris* L.): evidence for multiple centers of domestication. *Econ. Bot*. 40: 451-468.

Goodman, R. M., H. Haupli, A. Crossway and V. C. Knauf. 1987. Gene transfer in crop improvement. *Science*. 236: 48-54.

_____ and C. W. Stuber. 1983. Races of maize. VI. Isozyme variation among races of maize in Bolivia. *Maydica*. 28: 169-187.

Graner, A., H. Siedler, A. Jahoor, R. G. Herrmann and G. Wenzel. 1990. Assessment of the degree and the type of restriction fragment length polymorphism in barley (*Hordeum vulgare*). *Theor. Appl. Genet*. 80: 826-832.

Grun, P. 1990. The evolution of cultivated potatoes. *Econ. Bot*. 44(3 Supplement): 39-55.

Hamrick, J. L. and R. W. Allard. 1972. Microgeographical variation in allozyme frequencies in *Avena barbata*. *Proc. Nat. Acad. Sci.* 69: 2100-2104.

Hamrick, J. L. and M. J. W. Godt. 1990. Allozyme diversity in plant species. Pp. 43-63. In: A. H. D. Brown, M. T. Clegg, A. L. Kahler and B. S. Weir (eds.), *Plant population genetics, breeding and genetic resources*, Sinauer Associates Inc., Sunderland, Massachusetts, USA.

Hamrick, J. L. and L. R. Holden. 1979. Influence of microhabitat heterogeneity on gene frequency distribution and gametic phase disequilibrium in *Avena barbata*. *Evolution* 33: 521-533.

Hamrick, J. L., Y. B. Linhart and J. B. Mitton. 1979. Relationships between life history characteristics and electrophoretically detectable genetic variation in plants. *Ann. Rev. Ecol. Syst.* 10: 173-200.

Harlan, J. R. 1976. Genetic resources in wild relatives of crops. *Crop Sci.* 16: 329-333.

_____. 1978. Sources of genetic defense. *Ann. N. Y. Acad. Sci.* 287: 345-356.

_____. 1984. Gene centers and gene utilization in American agriculture. Pp. 111-129. In: *Plant genetic resources - a conservation imperative*. Westview Press, Boulder, Colorado.

_____ and J. M. J. de Wet. 1971. Toward a rational classification of cultivated plants. *Taxon* 20: 509-517.

Hawkes, J. G. 1966. Modern taxonomic work on the *Solanum* species of Mexico and adjacent countries. *Amer. Potato J.* 43: 81-103.

_____. 1978. Genetic poverty of the potato in Europe. Pp. 19-27. In: *Broadening the genetic base of crops*. PUDOC, Wageningen. Wageningen, The Netherlands.

_____. 1990. *The potato - Evolution, biodiversity and genetic resources*. Belhaven Press, London, United Kingdom, 259p.

_____ and Hjerting, J. P. 1989. *The potatoes of Bolivia: their breeding value and evolutionary relationships*. Clarendon Press, Oxford, United Kingdom, 472p.

Helentjaris, T., G. King, M. Slocum, C. Siedenstrang and S. Wegman. 1985. Restriction fragment polymorphisms as probes for plant diversity and their development as tools for applied plant breeding. *Plant Mol. Biol.* 5: 109-118.

Hosaka, K. and R. E. Hanneman. 1988. Origin of chloroplast DNA diversity in Andean potatoes. *Theor. Appl. Genet.* 76: 333-340.

Iwanaga, M., R. Freyre and K. Watanabe. 1991. Breaking the crossability barriers between disomic tetraploid *Solanum acaule* and tetrasomic tetraploid *S. tuberosum. Euphytica* 52: 183-191.

Jackson, M. T. and B. V. Ford-Lloyd. 1990. Plant genetic resources - A perspective. Pp. 1-17. In: M. T. Jackson and B. V. Ford-Lloyd (eds.), *Climatic change and plant genetic resources.* Belhaven Press, London, United Kingdom.

Jain, S. K., L. Wu and K. R. Vaidya. 1980. Levels of morphological and allozyme variation in Indian amaranths: a striking contrast. *J. Hered.* 71: 283-285.

Jana, S. and L. N. Pietrzak. 1988. Comparative assessment of genetic diversity in wild and primitive cultivated barley in a center of diversity. *Genetics.* 119: 981-990.

Kahler, A. L., R. W. Allard, M. Krzakowa, C. F. Wehrhahn and E. Nevo. 1980. Associations between isozyme phenotypes and environment in the slender wild oat (*Avena barbata*) in Israel. *Theor. Appl. Genet.* 56: 31-47.

Khairallah, M. M., B. B. Sears and M. W. Adams. 1991. Mitochondrial restriction fragment length polymorphisms in wild *Phaseolus vulgaris* L. *Annu. Rept. Bean Improvement Coop.* 34: (in press).

Khush, G. S. and K. C. Ling. 1974. Inheritance of resistance to grassy stunt virus and its vector in rice. *J. Hered.* 65: 134-136.

Koenig, R. and P. Gepts. 1989. Allozyme diversity in wild *Phaseolus vulgaris*: further evidence for two major centers of genetic diversity. *Theor. Appl. Genet.* 78: 809-817.

Ladizinsky, G. 1985. Founder effect in crop-plant evolution. *Econ. Bot.* 39: 191-199.

Lee, D., T. H. N. Ellis, L. Turner, R. P. Hellens and W. G. Cleary. 1990. A copia- like element in *Pisum* demonstrates the uses of dispersed repeated sequences in genetic analysis. *Plant Mol. Biol.* 15: 707-722.

Levy, A. A. and M. Feldman. 1987. Increase in grain protein

percentage in high- yielding common wheat breeding lines
by genes from wild tetraploid wheat. *Euphytica*. 36:
353-359.

Maquet, A., A. Gutierrez and D. G. Debouck. 1990. Further
biochemical evidence for the existence of two gene pools
in lima beans. *Annu. Rept. Bean Improvement Coop.* 33:
128-129.

Marshall, D. R. 1990. Crop genetic resources: current and
emerging issues. Pp. 367-388. In: A. H. D. Brown, M. T.
Clegg, A. L. Kahler and B. S. Weir (eds.), *Plant population genetics, breeding and genetic resources*. Sinauer
Associates Inc., Sunderland, Massachusetts, USA.

McLeod, M. J., S. I. Guttman, W. H. Eshbaugh and R. E. Rayle.
1983. An electrophoretic study of evolution in *Capsicum*
(Solanaceae). *Evolution*. 37: 562-574.

Menancio, D. I., A. G. Hepburn and T. Hymowitz. 1990.
Restriction fragment length polymorphism (RFLP) of wild
perennial relatives of soybean. *Theor. Appl. Genet.* 79:
235-240.

Miller, J. C. and S. D. Tanksley. 1990. RFLP analysis of phylogenetic relationships and genetic variation in the genus
Lycopersicon. Theor. Appl. Genet. 80: 437-448.

Mukherjee, S. K. 1985. *Systematic and ecogeographic studies of
crop genepools: 1. Mangifera L.* International Board for
Plant Genetic Resources, Rome, Italy. 86p.

Myers, N. 1983. *A wealth of wild species - Storehouse for
human welfare*. Westview Press Inc., Boulder, Colorado.
272p.

Nabhan, G. P. 1990. *Wild Phaseolus ecogeography in the Sierra
Madre Occidental, Mexico: areographic techniques for tar-
geting and conserving species diversity*. International
Board for Plant Genetic Resources, Rome, Italy. 35p.

National Research Council. 1972. *Genetic vulnerability of
major crops*. National Academy of Sciences, Washington,
USA, p.307.

Nevo, E., A. Beiles, D. Kaplan, E. M. Golenberg, L. Olsvig-
Whittaker and Z. Naveh. 1986a. Natural selection of allozyme polymorphisms: a microsite test revealing ecological
genetic differentiation in wild barley. *Evolution* 40:

13-20.

_____, _____ and Zohary, D. 1986b. Genetic resources of wild
    barley in the Near East: structure, evolution and applica-
    tion in breeding. *Biol. J. Linnean Soc*. 27: 355-380.

_____, A. H. D. Brown, D. Zohary, N. Storch and A. Beiles.
    1981. Microgeographic edaphic differentiation in allozyme
    polymorphisms of wild barley (*Hordeum spontaneum*, Poac-
    eae). *Pl. Syst. Evol*. 138: 287-292.

_____, D. Zohary, A. H. D. Brown and M. Haber. 1979. Genetic
    diversity and environmental associations of wild barley,
    Hordeum spontaneum, in Israel. *Evolution* 33(3): 815-833.

Nybom, H. 1990. Genetic variation in ornamental apple trees
    and their seedlings (*Malus*, Rosaceae) revealed by DNA
    'fingerprinting' with the M13 repeat probe. *Hereditas*.
    113: 17-28.

Oldfield, M. L. 1989. *The value of conserving genetic
    resources*. Sinauer Associates Inc., Sunderland, Massachu-
    setts, USA. 379p.

Osborn, T. C., D. C. Alexander, S. S. M. Sun, C. Cardona and
    F. A. Bliss. 1988. Insecticidal activity and lectin
    homology of arcelin seed protein. *Science*. 240: 207-210.

_____, T. Blake, P. Gepts and F. A. Bliss. 1986. Bean arcelin.
    2. Genetic variation, inheritance and linkage relation-
    ships of a novel seed protein of *Phaseolus vulgaris* L..
    Theor. Appl. Genet. 71: 847-855.

Osborne, D. J. 1982. Deoxyribonucleic acid integrity and
    repair in seed germination: the importance in viability
    and survival. Pp. 435-464. In: A. A. Kahn (ed.), *The phy-
    siology and biochemistry of seed development, dormancy and
    germination*. Elsevier, Amsterdam, The Netherlands.

Palmer, J. D., R. A. Jorgensen and W. F. Thompson. 1985. Chlo-
    roplast DNA variation and evolution in *Pisum*: patterns of
    change and phylogenetic analysis. *Genetics*. 109: 195-213.

Pickersgill, B. 1971. Relationships between weedy and culti-
    vated forms in some species of chili peppers (genus *Capsi-
    cum*). *Evolution*. 25: 683-691.

_____. 1984. Migrations of chili peppers, *Capsicum* spp., in
    the Americas. Pp. 105-123. In: D. Stone (ed.), *Pre-
    Columbian Plant Migration*. Harvard, Massachusetts.

Prescott-Allen, R. and C. Prescott-Allen. 1983. *Genes from the wild - Using wild genetic resources for food and raw materials*. Earthscan, International Institute for Environment and Development, London. 101p.

Pyle, M. M. and R. P. Adams. 1989. *In situ* preservation of DNA in plant specimens. *Taxon*. 38: 576-581.

Rick, C. M. 1976. Natural variability in wild species of *Lycopersicon* and its bearing on tomato breeding. *Genet. Agr*. 30: 249-259.

_____. 1978. Potential improvement of tomatoes by controlled introgression of genes from wild species. Pp. 167-173. In: *Broadening the genetic base of crops*. PUDOC. Wageningen, The Netherlands.

_____. 1988. Tomato-like nightshades: affinities, autoecology, and breeders' opportunities. *Econ. Bot*. 42: 145-154.

_____, J. W. DeVerna, R. T. Chetelat and M. A. Stevens. 1987. Potential contributions of wide crosses to improvement of processing tomatoes. *Acta Hort*. 200: 45-55.

_____ and Fobes, J. F. 1975. Allozyme variation in the cultivated tomato and closely related species. *Bull. Torrey Bot. Club*. 102: 376-384.

Rogers, S. O. and A. J. Bendich. 1985. Extraction of DNA from milligram amounts of fresh, herbarium and mummified plant tissues. *Plant Mol. Biol*. 5: 69-76.

Ross, H. 1978. Wild species and primitive cultivars as ancestors of potato varieties. Pp. 237-245. In: *Broadening the genetic base of crops*. PUDOC, Wageningen. Wageningen, The Netherlands.

Rowe, P. R. 1969. Nature, distribution, and use of diversity in the tuber-bearing *Solanum* species. *Econ. Bot*. 23: 330-338.

Saghai-Maroof, M. A., K. M. Soliman, R. A. Jorgensen and R. W. Allard. 1984. Ribosomal DNA spacer-length polymorphisms in barley: Mendelian inheritance, chromosomal location, and population dynamics. *Proc. Natl. Acad. Sci. USA*. 81: 8014-8018.

Schinkel, C. and P. Gepts. 1988. Phaseolin diversity in the tepary bean, *Phaseolus acutifolius* A. Gray. *Plant Breeding* 101: 292-301.

Second, G. 1982. Origin of the genetic diversity of cultivated
    rice (*Oryza* spp.): study of the polymorphism scored at 40
    isozyme loci. *Jpn. J. Genet.* 57: 25-57.
_____. 1985. Evolutionary relationships in the *sativa* group of
    Oryza based on isozyme data. *Genet. Sel. Evol.* 17: 89-114.
Shattuck-Eidens, D. M., R. N. Bell, S. L. Neuhausen and T.
    Helentjaris. 1990. DNA sequence variation within maize and
    melon: observations from polymerase chain reaction ampli-
    fication and direct sequencing. *Genetics.* 126: 207-217.
Simmonds, N. W. 1979. *Principles of crop improvement.* Long-
    man, London, United Kingdom. 408p.
Skerman, P. J., D. G. Cameron and F. Riveros. 1988. The tropi-
    cal pasture legumes. Pp. 194-487. In: Skerman, P. J., D.
    G. Cameron and F. Riveros (eds.), *Tropical Forage Legumes.*
    Food and Agriculture Organization of the United Nations,
    Rome, Italy.
Song, K. M., T. C. Osborn and P. H. Williams. 1988. *Brassica*
    taxonomy based on nuclear restriction fragment length
    polymorphisms (RFLPs). 2. Preliminary analysis of subspe-
    cies within *B. rapa* (syn. *campestris*) and *B. oleracea.*
    Theor. Appl. Genet: 76: 593-600.
_____, _____ and _____. 1990. *Brassica* taxonomy based on
    nuclear restriction fragment length polymorphisms (RFLPs).
    3. Genome relationships in *Brassica* and related genera
    and the origin of *B. oleracea* and *B. rapa* (syn. *campes-
    tris*). *Theor. Appl. Genet.* 79: 497-506.
Sprecher, S. L. 1988. *Allozyme differentiation between gene
    pools in common bean (Phaseolus vulgaris L.), with special
    reference to Malawian germplasm.* PhD. Michigan State Uni-
    versity, East Lansing, Michigan, USA. 207p.
Stalker, H. T. 1980. Utilization of wild species for crop
    improvement. *Adv. Agron.* 33: 111-147.
Stebbins, G. L. 1990. Introduction. Pp. 1-4. In: D. E. Soltis
    and P. S. Soltis (eds.), *Isozymes in plant biology.* Dios-
    corides Press, Portland, Oregon, USA.
Talbert, L. E., G. M. Magyar, M. Lavin, T. K. Blake, and S. L.
    Moylan. 1991. Molecular evidence for the origin of the
    S-derived genomes of polyploid *Triticum* species. *Amer. J.
    Bot.* 78: 340-349.

Toro, O., J. Tohme and D. G. Debouck. 1990. *Wild bean (Phaseolus vulgaris L.): description and distribution.* International Board for Plant Genetic Resources and Centro Internacional de Agricultura Tropical, Cali, Colombia. 106p.

Vargas, J., J. Tohme and D. G. Debouck. 1990. Common bean domestication in the southern Andes. *Annu. Rept. Bean Improvement Coop.* 33: 104-105.

Warnock, S. J. 1988. A review of taxonomy and phylogeny of the genus *Lycopersicon*. *HortScience*. 23: 669-673.

Weising, K., J. Schell and G. Kahl. 1988. Foreign genes in plants: transfer, structure, expression, and applications. *Annu. Rev. Genet.* 22: 421-477.

Wellhausen, E. J., L. M. Roberts and E. Hernandez Xolocotzi. 1952. *Races of maize in Mexico. Their origin, characteristics and distribution.* Bussey Institution, Harvard University, Harvard, Massachusetts. 223p.

Zhang, Q., M. A. Saghai Maroof and R. W. Allard. 1990. Effects on adaptedness of variations in ribosomal DNA copy number in populations of wild barley (*Hordeum vulgare* ssp. *spontaneum*). *Proc. Natl. Acad. Sci. USA*. 87: 8741-8745.

Zohary, D. and M. Hopf. 1973. Domestication of pulses in the Old World. *Science*. 182: 887-894.

FREEZE-PRESERVATION OF EMBRYOGENIC *MUSA* SUSPENSION CULTURES

B. Panis, D. Dhed'a and R. Swennen

Laboratory of Tropical Crop Husbandry
Catholic University of Leuven
Kardinaal Mercierlaan 92
B-3001 Heverlee, Belgium

Summary - Embryogenic cell suspension cultures of two
Musa spp., the cooking banana cv. *Bluggoe* (ABB group) and the
wild BB diploid *Musa balbisiana*, have been preserved in liquid
nitrogen. The frozen suspension cultures were regenerated into
plants through somatic embryogenesis. This procedure seems to
be promising for the long-term preservation of *Musa* germplasm.

INTRODUCTION

By the 1960's it was realized that many wild species were
about to disappear and that the size of crop gene pools of
major economic importance was shrinking at an alarming rate.
However, it was not until the establishment of IBPGR (International Board for Plant Genetic Resources) in 1974 that a
worldwide program for the preservation of tropical and subtropical germplasm started. Crops were classified by
priority, the highest priority was given to crops which needed
the most urgent attention. Bananas and plantains were rated
among the 50 crops needing the most urgent action (Ford-Lloyd
and Jackson, 1986).

Banana and plantain (*Musa* spp.) are important staple food
crops in the humid and subhumid tropics. About 69 million
tons are produced annually (FAO, 1989), of which only 10% are
exported (INIBAP, 1988). They are mainly grown as food for
local consumption on small subsistence farms or in backyard
gardens. Many pests and diseases are threatening banana and
plantain production, among which are black sigatoka, a fungal
disease caused by *Mycosphaerella fijiensis*; bunchy top, a
virus disease causing plant distortion; nematodes; and banana
weevils (Persley and De Langhe, 1987). The seriousness of the
black sigatoka threat alarmed the international community and

resulted in the creation of INIBAP (International Network for
the Improvement of Banana and Plantain) in 1984.

Nearly 1000 banana cultivars are known to exist. The
majority of those of immediate importance for mankind produce
their fruits parthenocarpically and are sterile. These cultivars which are either diploid, triploid or tetraploid can thus
only be propagated and conserved vegetatively. However, wild
bananas which are diploid produce seeds in their fruits and
are thus propagated vegetatively and by seed. Most varieties
have not yet been systematically evaluated for their yield
potential or resistance to pests and diseases. This germplasm
needs thus to be urgently preserved.

The most convenient way to maintain plant germplasm is by
storing seeds. This is not feasible with recalcitrant seeds
and for the many banana and plantain cultivars which can only
be vegetatively propagated. Therefore field (ex *situ*) collections have been established. These collections demand large
inputs of labor and land, and are subject to diseases or other
hazards (Simmonds, 1979). An alternative proposition to the
field (ex *situ*) genebank that overcomes many of these problems
is the storage *in vitro*. Under these conditions, it is necessary, however, to reduce the growth rate to a minimum. In
Musa, minimal growth is achieved by maintaining shoot-tip cultures at 2000 lux and at 15±2°C necessitating one annual sub-
culturing only of shoot-tips (Banerjee and De Langhe, 1985;
Withers, 1984). The INIBAP *Musa* germplasm collection which
consists of nearly 1000 accessions is currently maintained
under such conditions at its transit center at the Laboratory
of Tropical Crop Husbandry, Leuven, Belgium (Figure 1). Yet
there are some drawbacks in maintaining germplasm *in vitro*: 1)
the occurrence of somaclonal variation (Vuylsteke *et al.*,
1988; Vuylsteke *et al.*, 1991), 2) the risk of losing accessions due to contamination or human error at every subcultur-
ing, 3) the high labor demand when dealing with such a large
collection.

The preservation of germplasm collections under cryopreserved conditions would be an attractive proposition for the
long term conservation. Storage in liquid nitrogen (-196°C)
seems to be the method of choice, because under these condi-

Figure 1   The *Musa* germplasm collection (about 1000 accessions) at the INIBAP Transit Center where meristems are preserved *in vitro* under limited growth conditions.

tions all chemical reactions in the cell have ceased due to
the low temperature and unavailability of liquid water.   Hence
no cell division, cell degeneration or genetic changes can
take place in time (Hauptmann and Widholm, 1982).

Research on the cryopreservation of *Musa* is very scanty.
Freezing experiments in *Musa* were initially carried out with
meristems (Banerjee, Panis, unpublished).   Whatever the freezing method, no survival was obtained when the samples were
subjected to temperatures below the freezing point of the cry-

oprotectant solution. Recently, Mora et al. (1991) reported
the survival of zygotic embryos of *Musa acuminata* sp. *burma-
niccoides* and *Musa balbisiana* after freezing in liquid
nitrogen. This protocol however only offers promise for seed
producing bananas. In this paper we propose an *in vitro* conservation methodology which is also applicable to bananas
which do not produce seed.

Usually six stages are distinguished in the cryopreservation process i.e. pregrowth, cryoprotection, freezing, storage, thawing and post-thaw phase.

PREGROWTH

A physically and chemically induced pregrowth phase is
often applied to increase freezing tolerance of tissue cultures. For example, addition of osmotically active compounds
like sorbitol and mannitol reduces the cellular water before
freezing, thus reducing the amount of water available for
lethal ice formation (Withers and Street, 1977).

CRYOPROTECTION

It is speculated that cryoprotectant solutions maintain
cellular water in a liquid state at low temperatures, promote
cell dehydration and membrane stabilization.

FREEZING, STORAGE AND THAWING

Slow and rapid freezing methods are distinguished in the
cryopreservation of plant tissue culture. The former is particularly suitable for suspension cultures and protoplasts
while the latter consisting of plunging the material directly
into liquid nitrogen is more successful for meristems, seeds
and pollen. Slow freezing allows cells to dehydrate. Ice
formation is initiated outside the cell, leading to a low
osmotic pressure of the solution. Water leaves the plant cell
in order to restore the equilibrium in vapor pressure between
cells and surrounding solution, thus reducing the amount of
freezable water. Rapid freezing also avoids ice damage: the
cooling takes place so rapidly that intracellular ice crystals
cannot grow to a damaging size before all of the intracellular
water has frozen.

POST-THAW PHASE

After thawing, sometimes the very toxic cryoprotectants are washed off. On the other hand a cryoprotectant like DMSO can play an important role as a free radical scavenger in the post-thaw stabilization of cryopreserved cultures of *Daucus carota* (Benson and Withers, 1987). It was also observed that washing often results in deplasmolysis injury (Withers and King, 1980).

MATERIALS AND METHODS

Recently, it became possible to initiate and to maintain embryogenic cell suspensions (ECS) derived from somatic tissue (Dhed'a *et al.*, 1991). These cultures seem to be the appropriate material for preservation in liquid nitrogen (Panis *et al.*, 1990).

Cell suspensions have been initiated from the cv. *Bluggoe* (*Musa* spp., ABB group) and the wild BB diploid, *Musa balbisiana*, using explants from the upper part of the meristematic shoot-tip. They were maintained in half-strength Murashige and Skoog (MS) medium supplemented with 5×10^{-6} M 2,4-dichloro-phenoxyacetic acid (2,4-D) and 10^{-6} M zeatin. These suspension cultures are very heterogeneous and contain large translucent cells as well as small dense cells. When subcultured at an interval of 3 weeks, suspension cultures become more uniform and consist of clusters of small, tightly packed cells with dense cytoplasm. These cells were round and characterized by a relatively large nucleus, very dense nucleolus, small multiple vacuoles and tiny starch and protein grains (Figure 2A). These characteristics indicate embryogenic competence (McWilliams *et al.*, 1974).

A pregrowth phase was carried out by subjecting *Musa* cells to osmotic stress by culturing in 6% mannitol for 2 or 7 days as commonly used (Withers and King, 1980). The most commonly applied cryoprotectant solutions were examined i.e. dimethylsulfoxide (DMSO), glycerol and proline in different concentrations (5, 7.5, 10 and 15 % v/v) and a mixture of 0.5 M DMSO, 0.5 M glycerol and 1 M sucrose (Withers and King, 1980). In each case an equal volume of chilled sterile medium containing the cryoprotectant at twice the concentration was

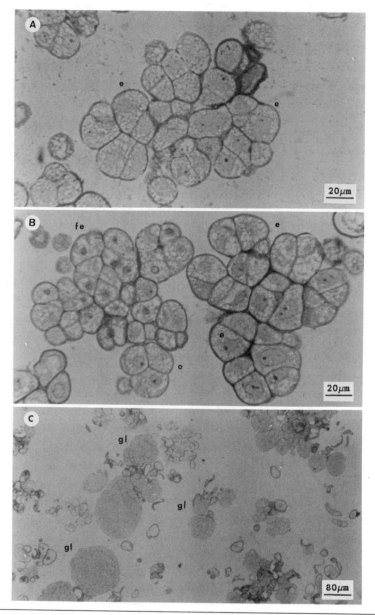

Figure 2. Embryogenic suspension culture of *Musa* (ABB) cv.
Bluggoe. A) Before freezing; b) Immediately after cryopreservation and C) cultured on semi-solid medium one month after
cryopreservation. e: normal embryogenic cells; fe: freeze-
injured embryogenic cells; gl: globules.

added gradually over 1 h to a suspension cooled in an ice
bath. The suspensions were taken two weeks after subcultur-
ing when they were in the exponential growth phase.

Samples of cryoprotected *Musa* cells were transferred to
cryotubes and cooled at a rate of $1°C min^{-1}$ to $-40°C$. This
temperature was maintained for 30 min, after which the tubes
were plunged into a Dewar flask, filled with liquid nitrogen.
After 30 min in liquid nitrogen, the samples were successfully
thawed by plunging the cryotubes in a water bath stirred at
$40°C$ for about 1 min until most of the ice had melted. As
such, recrystallization was avoided during the rewarming process.

The viability of frozen suspensions was microscopically
determined with the fluorescein diacetate (FDA) test and/or
the 2,3,5-triphenyl tetrazolium chloride (TTC) reduction test.
Success of cryopreservation was also assessed by regrowing
plantlets on semi-solid medium (Seitz and Reinhard, 1987).

RESULTS AND DISCUSSION

Our optimal cryopreservation protocol is equally simple
but differs in a number of ways to those for cell suspensions
of other plant species (Withers and King, 1980).

Preculturing the *Musa* cells in 6% mannitol did not affect
viability after freezing. Consequently, no special pregrowth
phase was included in our experiments.

Optimal cryoprotection was obtained by using 7.5 % DMSO
as the sole cryoprotectant. FDA tests confirmed viability but
estimates were consistently higher then in regrowth tests.
Discrepancy between the two viability tests increased dramati-
cally when glycerol was used as a cryoprotectant.

Regrowth rate of frozen suspensions on petri dishes was
increased from 60 to 90%, when ice-crystallization by seeding
took place at $-7.5°C$ during the slow freezing process. This
was realized by plunging suspension cultures in liquid
nitrogen for 3 sec as soon a temperature of $-7.5°C$ was
reached, after which they were further cooled to $-40°C$. It is
speculated that viability was increased due the avoidance of
excessive supercooling, thus allowing protective dehydration
of the cells.

The removal of the cryoprotectant solution directly after
thawing and its replacement by cryoprotectant free liquid
medium before transferring the cells to a semi-solid medium,
resulted in a complete loss of regrowth capacity, and the
cells becoming white. Direct transfer of cells to a liquid
medium, which subjects the cells to similar post-thaw wash
stresses, likewise resulted in growth failure. In contrast,
regrowth was achieved when thawed cells, still surrounded by
the cryoprotectant solution, were directly transferred to
semi-solid Murashige and Skoog medium containing 10^{-6} M benzy-
ladenine (BA) and 100 mg.1^{-1} myo-inosotol. Growth resumed
after 10 days. Four to 6 weeks later, a layer of small globules formed which were identified as somatic embryos (Figure
3A). After separation they were inoculated onto a hormone-
free medium where they developed into normal plantlets (Figure
3B). Alternatively, cells which resumed growth on the semi-
solid medium could be induced to multiply and form an embryogenic suspension culture if transferred to a liquid medium
after 3 weeks.

Histological observations on cryopreserved *Musa* suspension cultures revealed that only the most embryogenic cells
survived freezing. Cells that lost their viability exhibited
a granular cytoplasm with broken vacuoles. In addition, their
nucleus was very granular, irregular and contracted, probably
due to the excessive dehydration of the cells (Figure 3A).
Thirty days after thawing, globules which will develop later
into somatic embryos could be clearly distinguished (Figure
3B).

The freezing method which was first optimized for ECS of
Bluggoe (*Musa* spp. cv ABB group), gave similarly positive
results with the wild *Musa balbisiana* (BB group). While in
Bluggoe 90% of the ECS survived the freezing protocol, survival rate in *Musa balbisiana* amounted to 94%. In both
varieties, it was ascertained that only embryogenic cells
withstand freezing injury and that these cells regenerated
into plantlets through somatic embryogenesis, provided proper
culture conditions were met.

These encouraging results seem to suggest that the applicability of the freeze-preservation in liquid nitrogen of

Figure 3. Recovery of plantlets after cryopreservation.
A) Regrowth 6 weeks after thawing on semi-solid medium of
frozen cv. *Bluggoe* (*Musa* spp. cv. ABB) suspension culture
(left) and frozen *Musa balbisiana* suspension culture (right).
B) Mass of somatic embryos, originated from a frozen cell culture of cv. *Bluggoe*.   C) Plants regenerated from a frozen cell
culture of cv. *Bluggoe* grown in the greenhouse.

other *Musa* varieties depends on the development of ECS.   The
cryopreservation of ECS offers several comparative advantages

over other *Musa* tissues:

1) Regeneration of large numbers of plantlets is possible
through somatic embryogenesis, because cells with regeneration
potential are the ones which survive freezing. This contrasts
with the usual observation that the regeneration of plantlets
from a frozen suspension is often the limiting factor in the
cryopreservation of suspension cultures.

2) Frozen-thawed suspension cultures can directly be used
in genetic engineering programs. Our experience indicates
that ECS are the material of choice for the isolation of protoplasts.

3) Preservation of a *Musa* variety does not depend on its
seed producing capability as does the cryopreservation of
zygotic embryos.

Nevertheless, there are some limitations with the use of
suspension cultures, such as:

1) The isolation and establishment of embryogenic *Musa*
cultures is very time consuming (8 to 10 months) (Dhed'a *et
al.*, 1991).

2) Suspension cultures might be subject to mutation and
selection pressures before the freezing process is initiated.
This may lead to the production of abnormal progeny or even to
the loss of totipotency (Scowcroft, 1984).

It is therefore necessary to screen plants from frozen-
thawed ECS for their clonal uniformity. Note that the storage
in liquid nitrogen itself is the limiting factor, but the
stresses due to culture prior to freezing, cryoprotective
treatment, introduction to and recovery from storage can cause
genetic abnormalities. Consequently, plants obtained from
frozen ECS were grown and monitored under glasshouse conditions until they reached a height of 2 m (Figure 2C). Satis-
fying results prompted the evaluation of a larger population.
Therefore, a field has been established at the Onne station of
IITA (International Institute of Tropical Agriculture), which
is situated in southeastern Nigeria.

Vitrification could be an alternative way of freeze-
preserving plant tissues. Its advantage rests in the fact
that this method does not depend upon expensive equipment
required for the slow cooling of tissues. Vitrification is a

process whereby a solution is converted from the liquid phase directly into an amorphous glass phase, thus avoiding the formation of crystalline ice. This requires an ultra-rapid cooling rate and a sufficiently concentrated cryoprotective solution.

Experiments, similar to those carried out on nuclear cells of navel orange by Sakai *et al.* (1990) resulted in very poor survival of *Musa* cells. Only a few embryogenic clumps managed to regenerate. Most of the cells failed to recover due to the excessive plasmolysis of cells caused by the high osmotic pressure of the surrounding solution. Additional research on vitrification is thus needed before this storage method can also be considered for the freeze-preservation of *Musa* germplasm.

Acknowledgements

The Belgian Administration for Development Cooperation (BADC) and the International Network for the Improvement of Banana and Plantain (INIBAP) are gratefully acknowledged for their financial support and encouragements.

Literature Cited

Banerjee, N. and E. De Langhe. 1985. A tissue culture technique for rapid clonal propagation and storage under minimal growth conditions of *Musa* (Banana and plantain). *Plant Cell Rep.* 4: 351-354.

Benson, E. E. and L. A. Withers. 1987. Gas chromatographic analysis of volatile hydrocarbon production by cryopreserved plant tissue culture: A non-destructive method for assessing stability. *Cryo-Lett.* 8: 35-46.

Dhed'a D., F. Dumortier, B. Panis, D. Vuylsteke and E. De Langhe. 1991. Plant regeneration in cell suspension cultures of the cooking banana cv. *Bluggoe* (*Musa* spp, ABB group). Accepted for publication in *Fruits*.

FAO. 1989. *Production Year Book 1989*. Food and Agriculture Organization of the United Nations, Rome.

Ford-Lloyd, B. and M. Jackson. 1986. *Plant genetic resources: An introduction to their conservation and use.* Edward

Arnold Ltd., London, 149 pp.

Hauptmann, R. M. and J. M. Widholm. 1982. Cryostorage of cloned amino acid analog-resistant carrot and tobacco suspension cultures. *Plant Physiol.* 70: 30-34.

INIBAP. 1988. *Annual Report 1988.* International Network for the Improvement of Banana and Plantain, Montpellier.

McWilliams, A. A., S. M. Smith and H. E. Street. 1974. The origin and development of embryoids in suspension cultures of carrot (*Daucus carota*). *Ann. of Bot.* 38: 243-250.

Mora, A., A. Abdelnour and V. Villalobos. 1991. Cryopreservation of *Musa* zygotic embryos. *In: Biotechnology for tropical crop improvement in Latin America.* Abstracts of the fourth conference of the International Plant Biotechnology Network (IPBNet), San Jose, Costa Rica, Jan. 1991.

Panis, B., L. A. Withers and E. De Langhe. 1990. Cryopreservation of *Musa* suspension cultures and subsequent regeneration of plants. *Cryo-Lett.* 11: 337-350.

Persley, G. J. and E. A. De Langhe. 1987. *Banana and plantain breeding strategies*, ACIAR Proc. No. 21., Cairns, Australia, Oct. 1986. 187 pp.

Sakai, A., S. Kobayashi and I. Oiyama. 1990. Cryopreservation of nucellar cells of navel orange (*Citrus sinensis* Osb. var. *brasiliensis* Tanaka) by vitrification. *Plant Cell Rep.* 9: 30-33.

Scowcroft, W. R. 1984. Genetic stability of *in vitro* cultures. Consultant report. Rome: IBPGR, 62 pp.

Seitz, U. and E. Reinhard. 1987. Growth and ginsenoside patterns of cryopreserved *Panax ginseng* cell cultures. *J. Plant Physiol.* 131: 215-223.

Simmonds, N. W. 1979. Genetic conservation: An introductory discussion of needs and principles. Pp. 1-11. *In: IBPGR: Seed technology for genebanks*, IBPGR, Rome.

Vuylsteke, D., Swennen, R. and E. De Langhe. 1991. Somaclonal variation in plantains (*Musa* spp., AAB group) derived from shoot-tip culture. Accepted for publication in *Fruits*.

_____, G. F. Wilson and E. A. De Langhe. 1988. Phenotypic variation among *in vitro* propagated plantain (*Musa* spp.

cv. AAB). *Sci. Hortic.* 36: 70-88.

Withers, L. A. 1984. Germplasm conservation *in vitro*: present
state of research and its applications. Pp. 138-157. <u>In:</u>
J. H. W. Holden and J. T. Williams (eds.): *Crop genetic
resources: Conservation and evaluation.* George Allen and
Unwin Ltd., London.

_____ and P. J. King. 1980. A simple freezing unit and cryo-
preservation method for plant cell suspensions. *Cryo-
Lett.* 1: 213-220.

_____ and H. E. Street. 1977. Freeze-preservation of cultured
plant cells III. The pregrowth phase. *Physiol. Plant.* 39:
171-178.

CRYOCONSERVATION OF *MUSA* SPP AND ITS POTENTIAL FOR LONG-TERM STORAGE OF OTHER TROPICAL CROPS

Víctor M. Villalobos and Ana Abdelnour

Unidad de Biotechnología
Centro Agronómico Tropical de Investigación y Enseñanza
Apartado Postal 15, CATIE
Turrialba, Costa Rica

Summary - Cryoconservation of bananas and plantains was evaluated as an alternative technique for long-term germplasm preservation. Dehydrated zygotic embryos of *Musa acuminata* (AA) and *M. balbisiana* (BB) maintained in liquid nitrogen were successfully germinated after thawing. The average rate of survival was evaluated by the percentage of germinated embryos. Up to 70% of *M. balbisiana* and 74% of *M. acuminata* grew into complete plants. The simplicity and efficiency of the present developed technique, provides a promising alternative for long-term conservation of *Musas* germplasm. Preliminary results also obtained in coffee zygotic embryos support the idea that the same approach could be utilized for other tropical crops.

INTRODUCTION

As a result of the accelerated loss of germplasm, both in wild and cultivated tropical species, conservation of genetic diversity is of vital importance. International concern for the disappearance of plant germplasm has increased in recent years. The conservation of plant germplasm has become an important consideration for future crop improvement programs and food security essential to the overall sustainability of the tropical equilibrium.

Due to the growing concern for the preservation of endangered plant germplasm, conservation methods for plant genetic resources have been developed, including: storage of orthodox seeds, *in situ* and *ex situ* conservation, *in vitro* storage of vegetative material and more recently, cryoconservation for long-term storage (Stushnoff and Fear, 1985). All of these

methods provide options for gene storage; however, tropical
crops have not been considered with the same interest as temperate crops despite the fact that tropical species represent
the most depleted germplasm. To define a conservation
strategy for tropical crops, certain factors must be considered, such as the nature of the material, the objective and
scope of conservation, the genetic stability and the technique
to be used (Rublo, 1985; Frankel, 1970).

CURRENT SITUATION OF *MUSAS*

For the improvement of bananas and plantains, germplasm
availability is critical. It is well documented that these
species are threatened by pests, devastating diseases and
environmental stresses. Conventional methods of genetic
improvement of *Musas* are being applied, but they are time consuming and desirable genotypes with resistant characteristics
are not easily obtained.

For developing countries in the humid tropics (outside the
rice production areas), the food produced and consumed in
rural, and especially isolated areas consists to a very large
extent of roots, tubers, bananas and plantains. Today these
crops are the most important source of calories for the majority of rural people in all those areas where the population
does not depend on rice, maize or wheat (Table 1).

Table 1. Production levels of some major starchy staples
1988 (100 MT).

REGION	CASSAVA	YAM	SWEET POTATO	BANANAS AND PLANTAIN
Africa	80.9	23.2	11.9	23.3
America and the Caribbean	24.5	1.0	3.6	22.8
Asia	41.6	0.1	76.0	18.1
Other	0.2	0.4	1.1	1.7
TOTAL	147.2	24.7	92.6	65.9

At present, the area cultivated with bananas and plantain in tropical America and the Caribbean represents over 1.5 million ha. In Costa Rica alone, 1,800 ha of plantain are destined for exportation. Nevertheless, plantain cultivated for local consumption is an important priority and is cultivated in intercropped systems with annuals and perennial crops like cocoa. Estimation of plantain production for the local market in this situation is very difficult. A similar situation is found in Colombia where more than half of the cultivated area, calculated at 228,000 ha is used for plantain as a subsistence crop. Considering the importance of these crops for several developing countries and recognizing that food security in those countries is highly dependent on this species, the International Board for Plant Genetic Resources (IBPGR) recommended in 1975, the establishment of a conservation program for bananas and plantain as priority crops (IBPGR, 1975). Furthermore, the International Network for the Improvement of Bananas and Plantain (INIBAP) has more recently begun supporting the international effort for germplasm storage and utilization worldwide (INIBAP, 1989).

STORAGE TECHNIQUES OF *MUSAS*

To understand the conservation strategy for banana and plantain, one must realize that the genus *Musa* includes two groups of plants. The wild plants are diploids, produce seeds and are classified as two species: *Musa acuminata* Colla (AA) and *Musa balbisiana* Colla (BB). These species give origin to the cultivars, which are characterized by female sterility, parthenocarpy, and for many cultivars various levels of male sterility; thus they are clones maintained exclusively through vegetative propagation by suckers and can be diploid, triploid or tetraploid (Champion, 1978; De Langhe, 1986).

In the past, *Musa* germplasm was maintained in field collections by national programs, regional centers, or individual specialists, and they comprised a limited number of accessions easily assimilated into breeding programs. These collections did not represent the full range of variability, and according to Williams (1986), they were not adequately identified, evaluated or documented. These materials were not considered as

collections due to the lack of proper documentation. It is
recognized that information on each accession is essential for
a complete utilization of *Musa* germplasm, as well as for other
species.

At present, there are 43 collections of *Musa* in 33 countries, among them: Jamaica, Honduras, Philippines, India and
Cameroon. These collections could be viewed as complementary
to each other and they are maintained as field genebanks (Williams, 1986). The fact that conservation of germplasm in
field genebanks has a number of disadvantages, such as the
risk of pests, diseases, natural disasters, high costs, and
others is well documented. Besides, the low multiplication
rate of bananas and plantains by conventional methods (suckers) hinders the rapid dissemination of improved genotypes.
In addition, the long periods of quarantine that the vegetative propagules must withstand impedes germplasm exchange
(Chiarappa and Karpati, 1984; De Langhe, 1984 and Stover,
1977).

Seed storage of fertile diploids of *Musas* has been studied
as an alternative, but it was only recently that research
showed this to be a viable possibility (Williams, 1986). However, seed germination of diploids is considered highly dependent on the maturity of the fruit when harvested and the conditions to which the seeds are subjected before germination.
It seems that the germination rate is significantly different
between batches of seeds (Escalant and Teisson, 1987; Simmonds, 1959).

STORAGE OF *MUSAS* GERMPLASM BY SLOW *IN VITRO* GROWTH

The conventional methods for plant propagation enable multiplication rates high enough to allow rapid diffusion and
utilization of new improved cultivars. Recently, the ability
to grow plant material in aseptic cultures has opened new possibilities for plant manipulation, breeding, and germplasm
conservation (Escalant et al., 1991).

In vitro propagation of *Musa* using shoot tips was reported
in the early 1970's (Berg and Bustamante, 1974; Ma and Shii,
1972; 1974) and culture procedures have improved considerably
since it is now possible to produce large quantities of plants

as well as callus, somatic embryos and zygotic embryos in a
short period of time (Bakry, 1984; Cronauer and Krikorian,
1984, 1985, 1987a,b; Escalant and Teisson, 1987; Vuylsteke,
1989).

The implementation of tissue culture techniques has made
great contributions to *Musa* germplasm conservation. Besides
the advantages mentioned above, it is possible to obtain material free of pathogens, viruses and nematodes which together
with the small size of the propagules, facilitate germplasm
exchange (INIBAP, 1989). However, Frankel (1970), considered
that *in vitro* conditions cannot be utilized as storage alternatives unless some modifications are made to reduce the rate
growth of *in vitro* cultivated plants and to diminish the risks
of contamination and genetic variation.

Withers (1989) indicated that the term slow growth is used
for growth initiation or minimal growth which imply a modifi-
cation of the culture conditions as well as naturally slow
growing material.

Successful storage of *Musa* shoot tips under minimal growth
conditions for up to one year has been achieved by reducing
the temperature and modifying the medium with osmotics or
growth retardants (Bannerjee and De Langhe, 1985).

Even though slow growth procedures are used routinely in
many laboratories and international centers, research is now
focussed on finding new techniques to assure genetic stability
of stored material and to extend the *in vitro* transfer
frequency, minimizing the risk of losses and the high costs of
labor and maintenance (Panis *et al.*, 1990).

FREEZE-STORAGE OF *MUSA* GERMPLASM

According to Engelmann (1991), cryoconservation is, at
present, the only suitable alternative for long-term storage
of germplasm. It seems to be the safest alternative because
once the material is frozen at -196°C, all metabolic functions
stop. As a result, no cell division, cell deterioration, or
mutation occurs. Moreover, the material can be maintained in
this state, theoretically, for unlimited length of time.

The objective of the cryopreservation procedure is to
bring biological material from normal physiological tempera-

tures to ultralow temperatures and back again to normal temperature without damaging it (Withers, 1985). The procedure
comprises the choice and isolation of the material, pretreatment, freezing, storage, thawing and recovery phases, which
should be defined specifically for each species.

This technique could be applied to all kinds of cultures,
and despite the fact that unorganized material generally per-
forms better than organized material, the selection process
should be based on the intrinsic genetic stability of the system, shoots and embryos being the priority cultures (Withers,
1980; 1987).

During the preculture phase some degree of dehydration is
induced in the cells and tissues to avoid damage caused by ice
crystal formation during the freezing and thawing process.
The material can be cultured from a few minutes to a few days
in the presence of cryoprotective substances such as sucrose,
mannitol, sorbitol, dimethylsulfoxide (DMSO), glycerol, or
other high molecular weight substances. In addition to induc-
ing protective dehydration, cryoprotectants have been shown to
lower the temperature at which intracellular water freezes and
to avoid excessive, potentially toxic concentrations of
solutes inside the cells. They also have an important role in
membrane stabilization and in the protection of enzyme binding
sites from freezing injury (Benson, 1990; Finkle et al., 1985;
Rudolph and Crowe, 1985; Withers and King, 1975).

Freezing can be ultra-rapid by direct immersion in liquid
nitrogen. For slow freezing a programmable freezing apparatus
is necessary in order to obtain precise and reproducible
results. Storage in liquid nitrogen is recommended. The liquid and vapor phases are stable at -150°C, thus preventing
progressive deterioration which might result from ice recrystallization (Withers, 1985). As mentioned above, the material
can be stored for unlimited length of time without deterioration.

Exposure to natural radiation during storage should not be
a concern because the frequency of mutations would only be
hazardous after several thousand years (Ashwood-Smith, 1985;
Engelmann, 1991).

Thawing must be carried out at a determined rate to pre-

vent recrystallization of any intracellular ice present (Withers, 1987). In most cases this is achieved by immersing the
ampules containing the samples in a water bath at 40°C; however, some exceptions have been reported where slow warming
was necessary (Katano *et al.*, 1983; Madox *et al.*, 1983).

The recovery phase consists of the treatment to which the
material is exposed after thawing. It can be achieved by
either washing out or diluting potentially toxic cryoprotectants and culturing the material under the best possible conditions to ensure its recovery (Engelmann, 1991).

Even though cryopreservation has been applied to many
plant species, all seven *Musa* species and cultivars have
received little attention and all early attempts to cryopreserve them failed. However, recent reports show some positive
results (Panis *et al.*, 1990; Mora *et al.*, 1991). Panis and
co-workers (1990) were able to cryopreserve suspension cultures of *Musa*, ABB, cultivar "Matavia" (Buggoe) using 5% DMSO
as a cryoprotectant with a freezing rate of 1°C/min to -40°C,
followed by plunging the cultures into liquid nitrogen. To
increase survival, freezing was induced at -10°C by interrupt-
ing the slow cooling to plunge the cryotubes for 0.5 sec. in
liquid nitrogen. Approximately 64% re-growth of cells and
cell clumps was obtained from which plantlets could be developed.

Mora (1990), using zygotic embryos of *M. acuminata* subspe-
cies Burmmanicoide (AA) and *M. balbisiana* (BB), obtained up to
74% survival of the embryos and regrowth into complete plants
by dehydrating them in a laminar flow cabinet (Figure 1).

Protocols similar to those used with *Musa* have also been
tried with other monocots such as cell suspensions of *Oryza
sativa* (Sala *et al.*, 1979) and zygotic embryos of *Hevea brasiliensis*, *Vertchia merrillii* (Chin *et al.*, 1988), *Elaeis
guineensis* (Grout *et al.*, 1983) and *Cocos nucifera* (Withers
and King, 1980). The similarity between different tropical
species with respect to cryopreservation is an advantage in
terms of planning strategies for other tropical species which
are considered valuable and endangered.

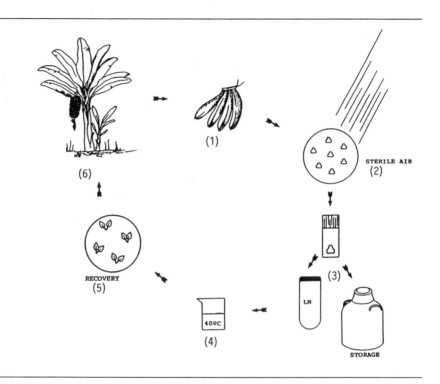

Figure 1.   Protocol for *Musa* zygotic embryos cryoconservation.
(1) Embryos extraction; (2) Embryos desiccation under sterile
air; (3) Freezing; (4) Thawing; (5) Embryos germination;
(6) Transfer to soil conditions.

From our preliminary studies (Mora *et al.*, 1991), it was
found that when high concentrations of sucrose, up to 0.75M,
were used as cryoprotectants, embryo viability of *M. acuminata*
decreased by up to 50%.   When DMSO was utilized at concentrations of 0, 5, and 10%, embryos of *M. balbisiana* incubated for
24 h were successfully germinated at 100%, but when they were
incubated in DMSO for 48 h, the response in viability was significantly reduced by up to 50%.   *M. acuminata* was even more
susceptible to DMSO during the tested incubation period.
       A key factor that counted for the success of cryocon-
servation of zygotic embryos of *Musas* was the desiccation

treatment provided by the laminar flow cabinet during different periods of time (Table 2). A 100% recovery was obtained when the desiccation method was applied to the embryos during two hours without freezing. In both species, a 100% recovery was obtained when embryos were subjected to a pretreatment of two hours of desiccation (8% humidity content).

Table 2. Water content expressed as percentage humidity of embryos of *Musa* dehydrated under a sterile flow.

Dehydration Period	Humidity (%)	
(hours)	*M. acuminata*	*M. balbisiana*
0	73	83
0.5	27	39
1.0	22	26
1.5	14	23
2.0	8	14
2.5	6	8

After freezing, *M. balbisiana* and *M. acuminata* showed a 70% and 74% survival respectively at two hours desiccation period (Table 3). It was clear that embryos without desiccation treatment did not survive. The use of sucrose in combination with desiccation treatment failed to enhance the germination rate.

CONCLUDING REMARKS
    Cryopreservation has been shown to be a viable alternative for germplasm conservation and despite the fact that it is not a routine technique for this purpose, the advantages that it presents over other methods of conservation makes it very attractive for the long-term germplasm conservation of recalcitrant and vegetatively propagated species. The simplicity

Table 3. Survival of zygotic embryos of *Musa* after freezing
in liquid nitrogen.

Dehydration Period (hours)	Survival (%)	
	M. acuminata	*M. balbisiana*
0	0	0
0.5	35	42
1.0	54	58
1.5	70	57
2.0	74	70
2.5	35	53

and efficiency of the *Musas* protocol used in our laboratory,
show its potential as a long-term conservation technique for
this important germplasm. Moreover, the possible application
of this simple protocol to other important tropical crops like
coffee, becomes an important alternative as an easy and
reliable method for long-term germplasm conservation.

Acknowledgements
    The authors gratefully acknowledge IBPGR for the grant to
CATIE for the cryopreservation of bananas and plantains No.
88/90. Also special thanks are given to Lissette Vega for her
assistance with this document.

Literature Cited

Ashwood-Smith, M. J. 1985. Genetic damage is not produced by
    normal cryopreservation involving either glycerol or
    dimethyl sulphoxide: a cautionary note, however, on pos-
    sible effects of dimethyl sulphoxide. *Cryobiology* 22:
    227-433.
Bakry, F. 1984. Choix du materiel a utilisar pour l' isolement
    de protoplastes de banajer (*Musa* spp.), Musacees. *Fruits*
    39: 449-452.
Bannerjee, N. and E. De Langhe. 1985. A tissue culture tech-

nique for rapid clonal propagation and Storage under minimal growth conditions of *Musa* (banana and plantain). *Plant Cell Reports* 4: 351-354.

Benson, E. 1990. Free radical damage in stored plant germplasm. Pp. 69-86. IBPGR. Rome.

Berg, L. A. and M. Bustamante. 1974. Heat treatment and meristem culture for the production of virus free bananas. *Phytopathology* 64: 320-322.

Champion, J. 1978. *El plátano*. 4th ed. Editorial Blume. España.

Chiarappa, L. and J. F. Karpati. 1984. Plant quarantine and genetic resources. In: J. H. Holden and J. T. Williams (eds.), *Crop genetic resources: Conservation and evaluation*. G. Allen and Unwin, London.

Chin, H. F., B. Krishnapillay and Z. C. Alang. 1988. Cryopreservation of *Veitchia* and *Howea* palm embryos: non-development of the haustorium. *Cryo-Letters* 9: 372-379.

Cronauer, S. and A. D. Krikorian. 1984. Rapid multiplication of bananas and plantains by *in vitro* shoot tip culture. *HortScience* 9: 325-328.

_____. 1985. Aseptic multiplication of banana from excised floral apices. *HortScience* 20: 770-771.

_____ and A. D. Krikorian. 1983. Multiplication of *Musa* from excised stem tips. *Ann. of Bot.* 53: 321-328.

_____ and _____. 1987a. Adventitious shoot production from calloid cultures of banana. *Plant Cell Reports* 6: 443-445.

_____ and _____. 1987b. Determinate floral buds of plantains (*Musa* AAB) as a site of adventitious shoot formation. *Ann. of Bot.* 61: 507-512.

De Langhe, E. 1984. The role of *in vitro* techniques in germplasm conservation. Pp. 131-137. In: J. H. Holden and J. T. Williams (eds.), *Crop genetic resources: Conservation and evaluation*. G. Allen and Unwin, London.

_____. 1986. Towards an international strategy for genetic improvement in the genus *Musa*. Pp. 19-28. In: G. J. Persley and E. A. De Langhe (eds.), *Banana and plantain breeding strategies*. ACIAR Proceedings, Australia.

Engelmann, F. 1991. Tropical plant germplasm conservation. In: Conference of the International Plant Biotechnology

Network. (4, 1991, San Jose, Costa Rica). Proceedings of
Biotechnology for Tropical Crop Improvement in Latin America, (in press).

Escalant, J. V. and C. Teisson. 1987. Comportements *in vitro*
de l'embryon isol du bananier (*Musa* species).

_____, V. Villalobos and M. Berthouly. 1991. La biotechnologie
apliquee aux trois cultures prioritaires pour le CATIE:
cafe, cacao et banane plantain. Pp. 7-17. In: Symposium
APES: "Les biotechnologies pour le Developpment de la Car-
aibe" (11, 1989, Fort de France, Martinique). (in press).

Frankel, O. H. 1970. Genetic conservation in perspective. Pp.
469-489. In: O. H. Frankel and E. Bennett (eds.), *Genetic
conservation in plants: Their exploration and conservation*. IBP Handbook N°11. F.A. Davis Co., Philadelphia.

Finkle, B. J., M. E. Zaval and J. M. Ulrich. 1985. Cryoprotective compounds in the viable freezing of plant tissues.
Pp. 75-113. In: K. K. Kartha (ed.), *Cryopreservation of
plant cells and organs*. CRC Press Inc., Florida.

Grout, B. W. W., E. Shelton and H. H. Pritchard. 1983. Orthodox behavior of oil palm seed and cryopreservation of the
excised embryo for genetic conservation. *Ann. of Bot.* 52:
381-384.

IBPGR. 1975. *The conservation of crop genetic Resources*. The
Whitefriars Press Ltd., London. 15 pp.

_____. 1989. *Partners in conservation. Plant genetic resources
and the CGIAR system*. Ed. CGIAR. Pp. 1-5.

INIBAP. 1989. *Annual Report*. France.

Jarret, R. L., W. Rodriguez and R. Fernandez. 1985. Evaluation, tissue culture propagation and dissemination of Saba
and Pelipita plantains in Costa Rica. *Sci. Hort.* 25:
137-147.

Katano, M., A. Ishihara and A. Sakai. 1983. Survival of dor-
mant apple shoot tips after immersion in liquid nitrogen.
HortScience 18: 707-708.

Ma, S. S. and C. T. Shii. 1972. *In vitro* formation of adventitious buds in banana shoot apex following decapitation.
China Horticulture 18: 135-142. English Summary.

_____ and _____. 1974. Growing banana plantlets from adventitious buds. *China Horticulture* 20: 6-12. English Summary.

Madox, A. D., F. G. Gonsalves and R. Shieles. 1983. Successful cryopreservation of suspension cultures of three *Nicotiana* species at the temperature of liquid nitrogen. *Plant Sci. Lett.* 28: 157-162.

Mora, A. 1990. Estudio del uso de la crioconservación de embriones cigóticos de *Musa balbisiana* y *Musa acuminata*. Tesis Mag. Sc. CATIE, Turrialba, Costa Rica. 109 Pp.

_____, A. Abdelnour and V. Villalobos. 1991. Cryopreservation of Zygotic Embryos of *Musa*. Conference of the International Plant Biotechnology Network. (4, 1991, San Jose, Costa Rica). Proceedings of Biotechnology for Tropical Crop Improvement in Latin America.

Panis, B. J., L. A. Withers and E. A. L. De Langhe. 1990. Cryopreservation of *Musa* suspension cultures and subsequent regeneration of plants. *Cryo-Letters* 11: 337-350.

Rubluo, A. 1985. Estrategias para la conservación del germoplasma vegetal *in vitro*. Pp. 35-53. In: M. Roberts and V. M. Loyola (eds.), *El cultivo de tejidos en Mexico*. Consejo Nacional de Ciencia y Tecnología.

Rudolph, A. S. and J. H. Crowe. 1985. Membrane stabilisation during freezing: The role of two natural cryoprotectants, trehalose and proline. *Cryobiology* 22: 367-377.

Sala, F., R. Cella and F. Rolo. 1979. Freeze-preservation of rice cells. *Plant Physiol.* 45: 170-176.

Simmonds, N. W. 1959. Experiments on the germination of banana seeds. *Trop. Agric. Trin.* 36: 259-274.

Stover, R. H. 1977. Banana (*Musa* spp.). Pp. 71-79. In: W. B. Hewritt and L. Chiarappa (eds.), *Plant health and quarantine in international transfer of genetic resources*. CRC Press, Boca Raton.

Stushnoff, C. and C. Fear. 1985. *The potential use of in vitro storage for temperate fruit germplasm: A status report*. IBPGR, Rome.

Vuylsteke, D. R. 1989. *Shoot-tip culture for the propagation, conservation and exchange of Musa germplasm*. IBPGR. Rome.

Williams, J. T. 1986. Banana and plantain germplasm conservational movement and needs of research. Pp. 177-180. In: G. J. Persley and E. A. De Langhe (eds.), *Banana and plantain breeding strategies*. ACIAR Proceedings, Australia.

Withers, L. A. 1980. *Tissue culture storage for genetic conservation*. IBPGR Technical Report. Rome. AGP:IBPGR/80/8
91 Pp.

_____. 1985. Cryopreservation of cultured plant cells and pro-
toplasts. Pp. 243-267. In: K. K. Kartha (ed.), *Cryopreservation of plant cells and organs*. CRC Press, Boca Raton.

_____. 1987. Long-term preservation of plant cells, tissues
and organs. *Oxford Surveys of Plant Molecular and Cell
Biology* 4: 221-272.

_____. 1989. *Musa* germplasm conservation. Prospects and prob-
lems of *in vitro* gene banks. IBPGR/INIBAP Musa Network
Meeting.

_____ and P. J. King. 1975. Proline. A novel cryoprotectant
for the freeze preservation of cultured cells of *Zea mays*
L. *Plant Physiology* 64: 675-678.

_____ and _____ . 1980. A simple freezing unit and routine
cryopreservation method for plant cell cultures. *Cryo-
Lett. 1: 213-220*.

CHROMOSOME VARIATION IN CALLUS CULTURE OF
GOSSYPIUM HIRSUTUM **L.**

Zheng Sijun

Department of Agronomy
Zhejiang Agricultural University
Hangzhou, Zhejiang, China

Summary - Embryogenic callus was induced with hypocotyls
of *Gossypium hirsutum* L. cv Henan 79, Simian 2, Lumian 6 and
Coker 312. Suspension cell cultures were established in four
genotypes. In a cytological study, 6-month cell cultures of
Coker 312 were analyzed and revealed to be variable in chromosome number and structure. The cell cultures showed a wide
range of chromosome numbers, varying from diploid to hexaploid
in euploid and from 16 to more than 78 in aneuploid lines.
The normal tetraploid cells were found only at 21.01%
frequency. Several mitotic abnormalities were found, such as
multi-nuclei and micronucleus in interphase; chromosome rings
and telophase bridges in the mitotic stage. Only 85.50% of
the cells of the suspension cultures were found to be normal
with a single interphase nucleus; 3.30% had two and 0.63% had
four interphase nuclei. The variation in chromosomes tended
to increase with the time in culture. The variability of
chromosomes in cell cultures seems to be one of the main reasons for failure of embryo development.

INTRODUCTION

Regeneration of *Gossypium hirsutum* L. has made it possible
for the use of somaclonal variation and gene manipulation in
improvement of existing breeding lines. Variation in chromosome number is one of several factors which can play an important role in the generation of somaclonal cells in higher
plants (Feher *et al.*, 1989; Larkin and Scowcroft, 1981; Pijnacker *et al.*, 1986 and Singh, 1986). Deviation from the normal chromosome set has been observed both in callus tissues
and in regenerated plants (D'Amato, 1985; Kononowicz *et al.*,
1990a,b). Several studies also report on a possible correla-

tion between the regeneration potential and the actual chromosome number (Murashige and Nakano, 1967; Torrey, 1967; Feher
et al., 1989). Bajaj and Gill (1985) reported hypocotyl-
derived callus cells of G. herbaceum and ovule-derived,
anther-derived callus cells of G. arboreum showed various
chromosome numbers, ranging from haploids to hexaploids, and
from high polyploidy to aneuploidy. Hybrid embryo-derived
callus cells of G. hirsutum x G. arboreum also showed a wide
range of chromosome numbers from less than 26 to more than 45.
The majority of the cells fell in the range of 26-41. No
polyploids were observed. Although some progress has been
achieved for tissue culture of cotton in recent years, it is
only limited in a few genotypes such as Coker 310, 312 etc.
(Finer, 1988; Trolinder and Goodin, 1988a,b). There was no
report of chromosomal variation in callus culture of G. hirsutum L. The genetic stability of plant cell cultures is important for the successful utilization of in vitro techniques in
plant breeding and germplasm preservation. The objective of
this work was to analyze chromosomal behavior of embryogenic
suspension lines of G. hirsutum L. cv Coker 312 for long term
culture and its effect on regeneration potential.

MATERIAL AND METHODS

   Culture conditions for aseptically-germinated seeds -
Seeds of G. hirsutum L. cv Henan 79, Simian 2, Lumian 6 and
Coker 312 were delinted with sulphuric acid and then surface-
sterilized for 30 minutes in 0.1% $HgCl_2$. The seeds were
rinsed 4 times in sterile water and placed on a medium containing 1/2 MS salt (Murashige and Skoog, 1962), 3% sucrose
and 0.8% agar (pH 5.8). Seedlings were obtained in 3 days
following germination at 28°C, under a 16:8 hours light: dark
photoperiod with a light intensity of 2000 lux.
   Callus induction - 3-5 mm hypocotyls were excised from
seedlings of four genotypes and were placed in S_1 medium for
callus induction, respectively. This medium contained MS
salts and B5 vitamins (Gamborg et al., 1968) (Table 1).

Table 1. Compositions of media used in cell cultures.

Constituent	Media (mg/l)		
	S_1	S_2	S_3
Salts	MS	MS	MS
Vitamins	B_5	B_5	B_5
KNO_3	-	-	1900
$AgNO_3$	-	5	-
Glutamine	-	-	2.19
Glucose	30000	30000	-
Sucrose	-	-	30000
2,4-D	0.05	0.01	-
KT	0.1	-	-
Agar	8000	8000	-

Conditions of callus initiation were the same as that for aseptically-germinated seedlings. Callus formation was observed at the cut edges of the explants in one week. All calli showed high growth rate and profuse callus development after one month.

　　Initiation of embryogenic callus - Callus produced in the S_1 medium were subcultured to S_2 medium (Table 1). In contrast to the S_1 medium, the S_2 medium contained 5 mg/l $AgNO_3$, 0.01 mg/l 2,4-D instead of 0.05 mg/l 2,4-D and 0.1 mg/l KT. Callus grew slowly in this medium. After 2-3 subcultures in the same medium, yellow, soft and friable embryogenic callus were produced. Some globular embryos were also observed.

　　Maintenance of embryogenic cell suspension cultures - Two months old, established, soft and friable embryogenic callus was used to initiate suspension cultures in liquid S_3 medium (Table 1). Compared with MS medium, it contained double concentration of KNO_3 and 15 mM glutamine, sucrose instead of glucose, and was devoid of auxin and cytokinin. Highly embryogenic suspension cell cultures were established using this media.

　　Cytological studies of cell cultures - Embryogenic cell suspension cultures have been established for 4 genotypes.

For the cytological analysis, 6-month cell cultures of Coker
312 were chosen. After the cell cultures were subcultured for
5-7 days, embryogenic callus cells were pretreated with saturated solution of p-dichlorobenzene (pDB) and α-bromo-
naphthalene for 3 hours at 25°C, then fixed in Carnoy fixative
solution for 2 hours. After fixation, the callus was transferred to 1N HCl for 15 minutes at a temperature of 60°C for
hydrolysis, then stained with modified carbol fushion for 2
hours. Slides were prepared by squashing. The preparations
were sealed with paraffin wax. Both chromosome counts and
cytological analysis were conducted using Olympus BH-2
research microscope.

RESULTS

Suspension 6-month cell cultures of Coker 312 contained
multi-nuclei in interphase of the cell cycle. Only 85.50%
of these suspension cultures were found to be normal with
single interphase nucleus, with 3.30% having two and 0.63%
having four interphase nuclei, respectively (Table 2). Cell

Table 2. Distribution of interphase nucleus and mitotic
nucleus in embryogenic callus cells of G. hirsutum L.

Cell cycle		Number of cells	% Total
1 nucleus	1350	85.50	
Interphase	2 nuclei	52	3.30
	4 nuclei	10	0.63
Mitotic		167	10.57
	Total	1579	100.00

cultures were observed to have good division ability with
a mitotic index of 10.45% and showed a high growth rate. It
was obvious that the size of normal nucleus was larger than
that of multi-nuclei (Figure 1).

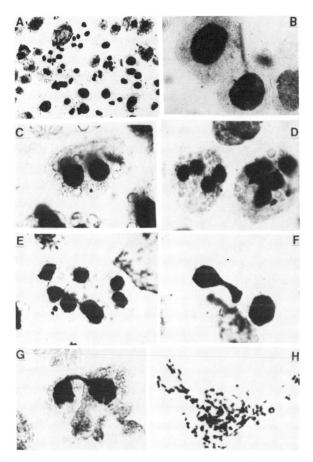

Figure 1. Mitotic abnormalities in embryogenic callus cells of
G. hirsutum L. cv Coker 312. A.) Cells with different sizes
of nuclei (260 x); B.) Cells with a single nucleus (870 x);
C.) Cell with two nuclei (1090 x); D.) Cells with four nuclei
(1090 x); E.) Micronuclei (1300 x); F.) Abnormal telophase
(1300 x); G.) Telophase bridge (1300 x); H.) Chromosome ring
at metaphase (870 x).

Various polyploid and aneuploid cells were observed (Table
3). They showed a wide range of chromosome numbers, varying
from diploid to hexaploid (Figure 2). The cells consisted of

normal tetraploid (2n = 52) cells (21.01%), diploids (10.95%),
triploids (7.40%), pentaploids (1.78%) and hexaploids (3.25%).
Only 46.76% of the cells were euploid, whereas the balance
were aneuploid for 338 analyzed cells. It was obvious that
most of the cells were diploid and tetraploid for the

Table 3. Variability of chromosome numbers in embryogenic callus cells of *G. hirsutum* L.

Chromosome number	Number of cells	% Total
16–25	19	5.62
26	37	10.95
27–38	84	24.85
39	25	7.40
40–51	72	21.30
52	71	21.01
53–64	3	0.89
65	6	1.78
66–77	2	0.59
78	11	3.25
>78	8	2.37
Total	338	100.00

euploids. The aneuploids showed a wide range of chromosome
number, varying from 16 to more than 78. However, the suspension cells with less than 26 and more than 78 chromosomes were
observed in relatively low frequencies of 5.62% and 2.37%,
respectively. The existence of several mitotic abnormalities
clearly appeared, such as micronucleus and multi nuclei formation in interphase stage, and chromosome ring, telophase
bridge and abnormal telophase in mitotic stage (Figure 1).
This phenomenon was also reported in tetraploid alfalfa by
Feher *et al.* (1989). It may be explained by the cytogenetic
instability that is often associated with cell cultures in
general.

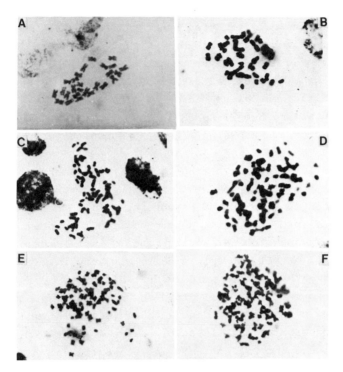

Figure 2.   Chromosomal variation in embryogenic callus cells
of *G. hirsutum* L. cv Coker 312.   A.) Diploid (2n = 26, 1300
x); B.) Triploid (2n = 39, 1300 x); C.) Tetraploid (2n = 52,
1090 x); D.) Pentaploid (2n = 65, 1300 x); E.) Hexaploid (2n =
78, 870 x); F.) Highly polyploid (2n > 78, 1090 x).

DISCUSSION

The success of any crop improvement program depends on the
extent of genetic variability in the base populations.   The
genetic instability of cell cultures is a well known phenomenon in most plant species.   Tissue cultures are known to be a
rich source of variability.   It is necessary to establish a
reliable program for plant regeneration.   As for cotton, callus is very easily induced and maintained with a high growth
rate since it is nonembryogenic.   Ethylene is produced by
plants and is known to have various effects on plant tissue
cultures.   Beyer (1976) reported the silver ion (Ag$^+$) was a

potent inhibitor of ethylene action and its incorporation into
tissue culture media produced beneficial effects on growth.
Similarly, silver nitrate (AgNO$_3$) improved regeneration of
callus cultures of Zea mays (Songstad et al., 1988) and Brassica oleracea (Williams et al., 1990). In the present paper,
the growth rate of callus decreased after subcultured to S$_2$
medium containing 5 mg/l AgNO$_3$. Finally, yellow and friable
embryogenic callus appeared after 2-3 subcultures in all four
genotypes.

The callus tissues in long-term culture underwent endomitosis, polyploid, chromosome loss, aneuploidy and other
genetic changes (Larkin and Scowcroft, 1981). The present
results reflected the same phenomenon from a cytological view
point. In suspension cell cultures of Coker 312, the
frequency of normal tetraploid cells was very low, only 21.01%
after six months. There were several different types of
mitotic abnormalities. Polyploid and highly polyploids may be
produced by endomitosis. Many authors (Cavallini and Natali,
1989; Kononowicz et al., 1990a,b) observed alteration of chromosome numbers and DNA contents in cell cultures. Chromosome
rings, telophase bridges and micronuclei may be the result of
chromosome structural changes. In addition, multi-nuclei and
chromosome loss were also observed in suspension cells.

The most important questions about somaclonal chromosome
variation are related to the possible effects of altered chromosome numbers on the regeneration potential. The major problem in present study is that complete plants were achieved
only within very low frequencies from somatic embryos. Most
of somatic embryos stopped further development. Some had
cotyledons but without roots. Others had hypocotyls and roots
but no bud formation. Vitrification was observed in the
somatic embryo development stage and some somatic embryos
evolved to callus-like structure (Figure 3). Polysomatic
embryos were observed in cell suspension cultures. The great
variability of chromosomes in cell cultures could be one of
the main reasons for the failure of embryo development.

To conclude, the genetic pool can be augmented by cell
cultures. The longer cells are in subcultures, the more vari-

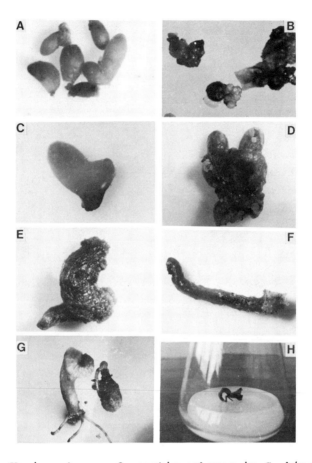

Figure 3. Various types of somatic embryos in *G. hirsutum* L.
cv Coker 312.   A.) Different development stages of somatic
embryos (9 x); B.) Somatic embryo evolved to callus-like
structure (3.5 x); C.) Abnormal somatic embryo (14 x); D., E.,
F.) Vitrification of somatic embryos (2.6 x , 3.5 x , 3.5 x ,
respectively); G., H.) Abnormal somatic embryos with hypocotyls and roots but no buds formation (2.6 x , 0.8 x , respectively).

ability is induced. This could be used to enhance the genetic
variability of the base populations. In addition, the genetic
stability of regeneration potential as well as the stability
of desirable traits that are selected under *in vitro* condi-

tions is of decisive importance, for example, in the utilization of artificial seeds. It is very important that the conversion rate from somatic embryos to whole plants is not significantly decreased while genetic variability is being increased as much as possible. The number of subcultures may be an important factor, because the variation of chromosomes tended to be increased with time in culture.

Acknowledgements - This research was supported by funds from the International Foundation for Science (C/1654-1). Thanks to Professors Ji Daofan and Xu Fuhua for assistance and Dr. Zhu Jun for help with computer techniques.

Literature Cited

Bajaj, Y. P. S. and M. S. Gill. 1985. *In vitro* induction of genetic variability in cotton (*Gossypium spp*). *Theor. Appl. Genet.* 70: 363-368.

Beyer, E. M., Jr. 1976. A potent inhibitor of ethylene action in plants. *Plant Physiol.* 58: 268- 271.

Cavallini, A. and L. Natali. 1989. Cytological analyses of in vitro somatic embryogenesis in *Brimeura amethystina salisb* (Liliaceae). *Plant Sci.* 62: 255-261.

D'Amato, F. 1985. Cytogenetics of plant cell and tissue cultures and their regenerates. CRC Crit. Rev. *Plant Sci.* 3: 73-112.

Feher, F., M. H. Tarczy, I. Bocsa and D. Dudits. 1989. Somaclonal chromosome variation in tetraploid alfalfa. *Plant Sci.* 60: 91-99.

Finer, J. J. 1988. Plant regeneration from somatic embryogenic suspension cultures of cotton (*Gossypium hirsutum* L.). *Plant Cell Reports.* 7: 399-402.

Gamborg, O. L., R. A. Miller and K. Ojima. 1968. Nutrient requirements of suspension cultures of soybean root cells. *Exp. Cell Res.* 50: 151-158.

Kononowicz, A. K., K. Floryanowicz-Czekalska, J. Clithero, A. Meyers, P. M. Hasegawa and R. A. Bressan. 1990a. Chromosome number and DNA content of tobacco cells adapted to NaCl. *Plant Cell Reports.* 8: 672-675.

_____, P. M. Hasegawa and R. A. Bressan. 1990b. Chromosome number and nuclear DNA content of plants regenerated from salt adapted plant cells. *Plant Cell Reports*. 8: 676-679.

Larkin, P. J. and W. R. Scowcroft. 1981. Somaclonal variation--a novel source of variability from cell cultures for plant improvement. *Theor. Appl. Genet.* 60: 197-214.

Murashige, T. and F. Skoog. 1962. A revised medium for rapid growth and bioassay with tobacco tissue cultures. *Physiol. Plant.* 15: 473-497.

_____ and R. Nakano. 1967. Chromosome complement as a determinant of the morphogenic potential of tobacco cells. *Am. J. Bot.* 54: 963.

Pijnacker, L. P., J. H. M. Hermelink and M. A. Ferwerda. 1986. Variability of DNA content and Karyotype in cell cultures of an interdihaploid *Sol tuberosum*. *Plant Cell Reports* 5: 43-46.

Singh, R. J. 1986. Chromosomal variation in immature embryo derived calluses of barley (*Hordeum vulgare*). *Theor. Appl. Genet.* 72: 710- 716.

Songstad, D. D., D. R. Duncan and J. M. Widholm. 1988. Effect of 1-aminocyclopropane-1-carboxylic acid, silver nitrate and norbornadiene on plant regeneration. *Plant Cell Reports* 7: 262-265.

Torrey, J. G. 1967. Morphogenesis in relation to chromosome constitution in long term plant tissue cultures. *Physiol. Plant.* 20: 265-275.

Trolinder, N. L. and J. R. Goodin. 1988a. Somatic embryogenesis in cotton (*Gossypium*). I. Effect of source of explant and hormone regime. *Plant Cell, Tissue and Organ Culture* 12: 31-42.

_____ and _____. 1988b. Somatic embryogenesis in cotton (*Gossypium*). II. Requirements for embryo development and plant regeneration. *Plant Cell, Tissue and Organ Culture* 12: 43-53.

Williams, J., D. A. C. Pink and N. L. Biddington. 1990. Effect of silver nitrate on long-term culture and regeneration of callus from *Brassica oleracea* var. *gemmifera*. *Plant Cell, Tissue and Organ Culture* 21: 61-66.

DNA POLYMORPHISM IN WHEAT AND ITS RELATIVES

P. P. Strelchenko, V. G. Konarev, O. Y. Antonova,
E. V. Horeva and M. N. Lapteva

N. I. Vavilov Institute of Plant Industry
Leningrad, 190000, USSR

Summary - Repetitive DNA sequences are useful molecular
markers for studying plant genome evolution and species divergence. In this paper, we report the cloning of wheat DNA
BspRI restriction fragments in the plasmid vector pUC19. By
the methods of dot blot and southern blot hybridization clones
of highly repetitive DNA suitable for *Triticeae* genomic DNA
polymorphism analysis were selected. The possibilities and
prospects of molecular markers useful in preservation and
mobilization of plant genetic resources are discussed.

INTRODUCTION

The erosion of natural plant genetic resources which has
resulted from people's economic activity and the deterioration
of global ecological situations at the present time may lead
to the essential impoverishment of initial material for breeding of various agricultural crops. Expeditionary gathering
and preservation of plants in regional and national plant
resources collections can, to some extent, reduce the acute-
ness of this problem, but at the same time demand more and
more financial expenditures. Therefore, the necessity of
applying new approaches for preservation of valuable and
important genes and its associations including DNA banking and
a creation *in vitro* systems is of great demand. Now it is
very important also to create new methods for registration and
systematization of world plant genetic resources. Molecular
markers are being more intensively used in identification of
varieties, lines, hybrids of various crops, in analysis of
plant populations and in the registration of plant genetic
resources. Molecular markers have some advantages over morphological ones. They are inherited codominantly, they are
not subject to gene pleiotropy and cannot be influenced by

environment. Methods using molecular markers based on electrophoresis of plural and genetically polymorphic proteins,
and indeed all enzymes, prolamins of grains and storage globulins of dicotyledon seeds have been successfully developed and
applied in plant development.

   At present, molecular markers are being widely used in
plant breeding in combination with different methods of selection beginning with analysis of the initial material up to
testing of varieties and subsequent seed farming (Konarev,
1983). Molecular markers have acquired special significance
in the registration and documentation of the gene pool of cultivated plants and their wild relatives in the form for "protein formulas" for varieties, lines, populations and biotypes. This way a number of crops have been registered, for
example - wheat, barley, rye, triticale, oats, corn, gramineous grasses, sunflower, sugarbeet and some leguminous and
cruciferous vegetable crops (Konarev, 1988).

   The prospects for development of plant breeding, seed
farming and methodology of registration, preservation and
effective utilization of plant genetic resources are associated with a second method of molecular marking of genetic and
biological plant systems. It is based on using the specificity of nucleic acids determined by their primary structure.
This specificity is usually evaluated according to the homology of molecules or their fragments by means of hybridization
methods. Restriction fragment length polymorphism (RFLP) is
the most widely used among existing methods of DNA hybridization (Burr *et al.*, 1983; Tanksley, 1983; Beckmann and Soller,
1986).

   In order to study the genomes of wheat and its relatives
we applied DNA-marking. In this paper we report the data of
DNA restriction polymorphism analysis with use of highly
repetitive cloned sequences.

MATERIALS AND METHODS

   Plant material - Seeds of plants were obtained from the N.
I. Vavilov Institute of Plant Industry, Leningrad, USSR. DNA
was isolated from the purified nuclear fraction, which was
prepared by filtering of germ homogenate through a system of

sieves. Nuclei were lysed and incubated with proteinase K
(Kislev and Rubenstein, 1980). Purification of DNA from
lysate was carried out by the phenol method removing the RNA
in 2M NaCl and then precipitation of DNA in CTAB. Since we
cannot detect any distinctions between electrophoretic patterns of restricts of DNA isolated from nuclei and etiolated
seedlings, we subsequently used the preparates of total cellular DNA from seedlings (Wienand and Feix, 1980). The size of
DNA was determined by electrophoresis in 0.3% agarose gels
with tris-borate buffer. DNA sizes were greater than 50 kb.

   Cloning of DNA - Nuclear DNA of *Triticum aesivum*, var.
Chinese Spring was completely digested with BspRI and cloned
into the HindIII site of pUC19 by blunt-end ligation. The
ligation mixture was used to transform *E. coli* JM 83 (Maniatis
et al., 1982). Colonies with recombinant plasmids were identified from transformants on the agar medium with X-gal and
IPTG. Plasmid DNA was isolated by alkaline extraction (Birnboim, 1983) and purified on cesium chloride-ethidium bromide
gradients.

   DNA hybridization - Because no significant hybridization
of *Triticum* DNA was obtained with the cloning vector pUC19,
the entire recombinant plasmids were labelled by nick translation (Rigby *et al.*, 1977) using deoxycytidine P-triphosphate
(3000 Ci/mmol) from Amersham. Total cellular or plasmid DNA
(400 ng) was applied without suction in a grid pattern to a
Millipore nitrocellulose membrane and dot blot hybridization
was carried out as previously described (Membrane transfer and
detection methods, Amersham International, 1985). For Southern blotting, 5 µg samples of DNA of each accession were
digested singly to completion, using 25 units of the following
restriction enzymes: HindIII, SduI, XbaI and Eco24I (Enzymas,
USSR). Restriction fragments were fractionated by gel electrophoresis on 0.8% agarose, denatured and transferred onto
Millipore membranes (Southern, 1975). Lanes of molecular size
standards were included on each gel. Southern blot hybridization was performed as described for dot blot hybridization.

   Hyperfilm-βmax film was used for autoradiography. White-
light absorbance of autoradiograms was measured in the linear
range of exposure, using LKB Ultroscan laser densitometer.

RESULTS AND DISCUSSION

We noticed the digestion of total cellular DNA of wheat
and subsequent restriction fragments in agarose gel gave distinct bands in electrophoretic pattern. We supposed these
bands are the restricts of the main families of genomic DNA
repetitive sequences. The results of comparative electrophoretic analysis of total DNA digested by Eco24I in a number of
Triticum species are shown in Figure 1. The analysis indi-

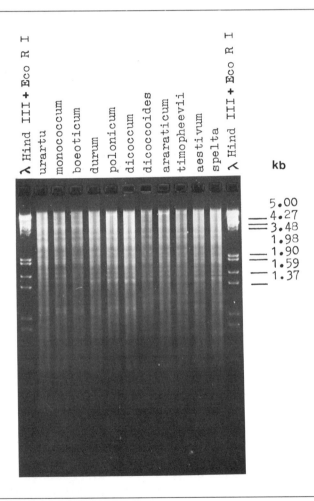

Figure 1. Electrophoresis of Eco24I digested genomic DNA of
some wheat species in 1.5 % agarose gel. HindIII and EcoRI
digested DNA of lambda phage were run as size markers.

cates a large similarity of electrophoretic patterns of these species. However, the pattern obtained for *T. urartu* was slightly different from other species. This results show that electrophoresis of total cellular DNA restriction may be used in preliminary analysis of DNA polymorphisms in *Triticeae*. It should be added that this method is useful to show differences on the intergeneric level (data not shown) or between remote species of a genus. Further improvement of this method will be combined with a detailed analysis of the molecular organization and evolution of repetitive DNA of individual taxa of the *Triticeae* for the subsequent introduction of nomenclature modifications.

Nuclear DNA of *T. aestivum* (var. Chinese Spring) was cloned into the vector plasmid pUC19 (Yanisch-Perron *et al.*, 1985) after digestion by BspRI restriction enzyme. In order to determine if DNA inserts belong to genomic DNA of wheat genome and to make a preliminary evaluation of their reiteration in the wheat genome, we conducted dot hybridization of recombinant plasmids of 136 randomly chosen clones with P^{32}-labelled total nuclear DNA of wheat. The results of dot hybridization indicate, that all clones gave a positive signal of different intensity, i.e. all inserts were fragments of wheat genomic DNA with different reiteration in the genome. Densitometric analysis of the auto-radiograms revealed good reproducibility of the dot hybridization results in three independent experiments, and approximately divide the clones into three groups according to their reiteration in the wheat genome. We determined that 18 of the 136 clones analyzed contained highly repetitive DNA and used them for further analysis. Subsequent analysis was carried out by the methods of dot blot and Southern blot hybridization with genomic DNA of different gramineous species. Dot blot analysis was used to select 2 clones from the 18 examined (pTa414 and pTa325), which were specific to the second genome of wheat polyploids and to *Aegilops* (Sitopsis section). The remaining 16 clones did not exhibit a genome specificity and they are present in all three wheat genomes and their supposed donors. Southern blot analysis showed, that not all of clones give distinct hybridization patterns and only some of then revealed DNA

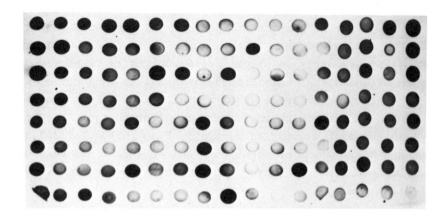

Figure 2. Dot hybridization of wheat P^{32}-labelled total
nucleic DNA with recombinant plasmids from the prepared
library of wheat nucleic DNA BspRI-fragments.

polymorphism in *Triticeae*. More interesting results were
obtained using 2 recombinant plasmids (pTa427 and pTa414).
The results of dot hybridization are shown in Figure 3. It
can be seen that these two clones represent families of DNA
repetitive sequences differed in the level of their specificity. Although clone pTa427 is specific to all analyzed
species of the tribe *Triticeae* and even insignificantly represented in nearest tribes of *Festucoides gramineous*, clone
pTa414 is specific only to subtribe *Triticeae*, particularly to
section Sitopsis of *Aegilops* and wheat polyploids. The
existence of such genome specific families of repetitive DNA
in plants has been reported by other authors (Flavell, 1982,
1983; Dvorak *et al.*, 1988; Zhao *et al.*, 1989; Gupta *et al.*,
1989; Vershinin *et al.*, 1990).

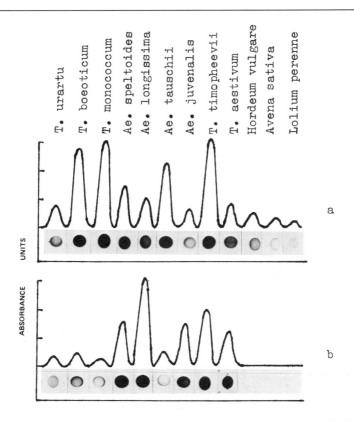

Figure 3. Dot hybridization of P[32]-labelled (a) pTa427 and
(b) pTa414 with genomic DNA of some *Poaceae* species. Densitometer traces of a single autoradiogram of dot blots on a
linear scale of absorbance (arbitrary units) are arranged with
Poaceae species. Dots were cut out for illustration. No
hybridization was detected in *Oryza sativa*, *Sorghum vulgare*
and *Phragmites australis* (hence not shown).

The results of Southern blot hybridization of *Triticeae*
genomic DNA with labelled plasmid pTa427 are shown in Figure
4.

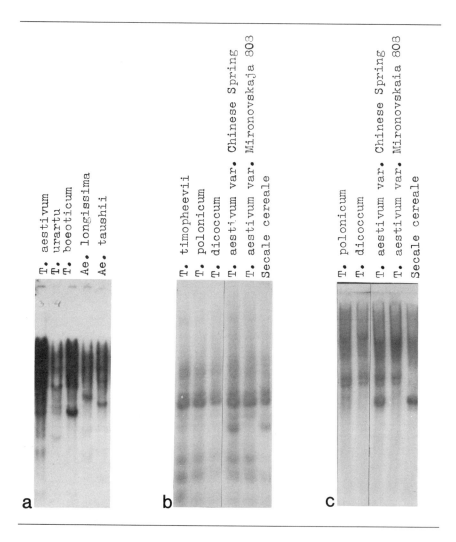

Figure 4. Southern blot hybridization of P³²-labelled of
pTa427 with some *Triticeae* species genomic DNA, digested with
(a) HindIII, (b) Sdu I and (c) Xba I.   The number of marking
lines are as in legend to Figure 1.

This plasmid contains a fragment from the most highly repetitive DNA family of the wheat genome. DNA polymorphisms in
wheats were determined using different restriction enzymes.

Some increase in the complexity of the hybridization pattern in polyploid wheats was observed in comparison with diploid ones (Figure 4a). It is interesting to note the differences in the blot patterns of two varieties: Chinese Spring and Mironovskaja 808. One of the main bands which was specific for Chinese Spring as well as for rye, was absent in Mironovskaja 808 (Figure 4b,c).

Southern hybridization with plasmid pTa 414 (Figure 5) shows its specificity for the second genome of wheat polyploids and *Aegilops* of section Sitopsis, which was also shown by dot hybridization (Figure 3b). This plasmid contains a part of a family with a highly repetitive DNA sequence, which probably has a tandem organization in genome and does not digest by HindIII (Figure 5a). We did find some enzymes which could digest this tandem. Eco24I revealed polymorphism among species of section Sitopsis of *Aegilops* (Figure 5b). As one can see, there are no differences between patterns of analyzed wheat species and *A. speltoides*, but the pattern obtained for *A. longissima* was different in its low molecular portion. In conclusion, restriction polymorphisms of genomic DNA can be used to study phylogenetic relationships among *Triticeae* as well as for the analyses of hybrids of various origins.

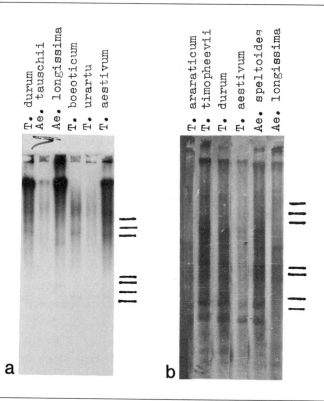

Figure 5.   Southern blot hybridization of P[32]-labelled of
pTa414 with some *Triticum* and *Aegilops* species genomic DNA,
digested with (a) HindIII and (b) Eco24I.   The number of marking lines are as in legend to Figure 1.

Literature Cited

Beckmann, J. S. and M. Soller. 1986. Restriction fragment
    length polymorphism in plant genetic improvement. *Oxford
    Surv. Plant Mol. Cell Biol.* 3: 197-250.

Birnboim, H. C. 1983. A rapid alkaline extraction method for
    the isolation of plasmid DNA. *Meth. Enzymol.* 100: 243-256.

Burr, B., S. V. Evola, F. Burr and J. S. Beckmann. 1983. The
    application of restriction fragment length polymorphism to

plant breeding. Pp. 45-49. In: J. Setlow and A. Hollander
(eds.) *Genetic engineering*. Plenum Press, New York.

Dvorak, J., P. E. McGuire and B. Cassidy. 1988. Apparent
sources of the A genomes of wheats inferred from polymorphism in abundance and restriction fragment length of
repeated nucleotide sequences. *Genome* 30: 680-689.

Flavell, R. 1982. Sequence amplification, deletion and rear-
rangement: major sources of variation during species divergence. Pp. 301-323. In: G. A. Dover and R. B. Flavell
(eds.), *Genome evolution*. Academic Press. London.

_____. 1983. Chromosomal variation at the molecular level in
crop plants. Pp. 15-25. In: L. D. Owens (ed.), *Genetic
engineering: Applications to agriculture*. Rowman and
Allanheld, Totowa, N.J.

Gupta, P. K., G. Fedrak, S. J. Molnar and R. Wheatcroft. 1989.
Distribution of a Secale cereale DNA repeat sequence among
25 *Hordeum* species. *Theor. Appl. Genet.* 32: 383-388.

Kislev, N. and I. Rubenstein. 1980. Utility of ethidium bromide in the extraction from whole plant of high molecular
weight maize DNA. *Plant Physiol.* 66: 1140-1143.

Konarev, V. G. 1983. Plant proteins as genetic markers. Moscow. Kolos. (In Russian.).

_____. 1988. Proteins in cultivar identification. Pp. 9-14.
In: V. G. Konarev and I. P. Gavriljuk (eds.), *Biochemical
identification of varieties*. Leningrad, USSR.

Maniatis, T., E. F. Fritsch and J. Sambrook. 1982. *Molecular
cloning. A laboratory manual*. Cold Spring Harbor Laboratory, Cold Spring Harbor, NY.

Rigby, P. W. J., M. Dieckmann, C. Rhodes and P. Berg. 1977.
Labelling deoxyribonucleic acid to high specific activity
in vitro by nick translation with DNA polymerase I. *J.
Mol. Biol.* 113: 237-251.

Southern, E. M. 1975. Detection of specific sequences among
DNA fragments separated by gel electrophoresis. *J. Mol.
Biol.* 98: 503-517.

Tanksley, S. D. 1983. Molecular markers in plant breeding.
Plant Mol. Biol. Rep 1: 3-8.

Vershinin, A. V., E. A. Salina, V. V. Solovyov and L. L. Timo-
feeva. 1990. Genomic organization, evolution, and struc-

tural peculiarities of highly repetitive DNA of *Hordeum vulgare*. *Genome*.33: 441-449.

Wienand, U. and G. Feix. 1980. Zein specififc restriction enzyme fragments of maize DNA. *FEBS Lett*. 116: 14-16.

Yanisch-Perron, C., J. Viera and J. Messing. 1985. Improved M 13 phage cloning vectors and host strains: nucleotide sequences of the M13 mp18 and pUC19 vectors. *Gene* 33: 103-119.

Zhao, X., T. Wu, Y. Xie and R. Wu. 1989. Genome-specific repetitive sequences in the genus *Oryza*. *Theor. Appl. Genet*. 78: 201-209.

IN SITU CONSERVATION OF PLANTS WITH POTENTIAL ECONOMIC VALUE

Chaia Clara Heyn

Department of Botany, The Hebrew University
Jerusalem 91904, Israel

and
Miriam Waldman

National Council for Research and Development
Ministry of Science and Technology
Jerusalem, Israel

Summary - *In situ* conservation seems the only method of conservation by which the genetic diversity as well as the dynamics within populations can be secured. A long-term multidisciplinary pilot study was carried out in a natural habitat of wild emmer wheat, *Triticum dicoccoides* (Koern.) Aarons. [*T. turgidum* L. var. *dicoccoides* (Koern.) Bowden] in Israel with the aim of developing an applicable methodology of *in situ* conservation. Though the resources in scientific manpower and land needed for such studies and for conservation may be large, it is proposed that *in situ* studies of genetic diversity should be made in populations of additional species, taking into account differences in breeding systems (selfers vers. outcrossers), taxonomic constraints (grasses vers. legumes) and environmental variation in different years in more than a single habitat. These studies will make it possible to arrive at conclusions concerning the feasibility of *in situ* conservation of certain species and develop a methodology for making decisions on the optimal attributes and the management of conserved areas. Being mainly concerned with the understanding of genetic diversity of the investigated species, *in situ* conservation studies could provide also an optimal model for sampling of populations for *ex situ* conservation of germplasm.

235

INTRODUCTION

The systematic conservation and utilization of the genetic diversity of wild progenitors of cultivated plants and other plants with potential economic value is limited by the available methodology of sampling. There is no doubt that conservation of germplasm by conventional methods has usually succeeded to secure only a small, and not always representative, fraction of the genetic resources. There is even a greater concern that this will happen in the future for plants still under investigation for their economic value. In fact, the majority of cultivated plants have been developed from only a small part of the genetic resources of the species which have usually been exploited partially and often haphazardly.

The widespread occurrence in Southwestern Asia of natural populations of plants which either have served as progenitors of some important cultivated plants or can be considered as being with potential economic value for the future has been repeatedly pointed out (e.g., Zoharya, 1983). In Israel, in which ecological conditions widely vary and in some areas the natural vegetation is endangered by the large human population density, several populations of progenitors of cultivated plants are included in nature reserves, which are established and maintained by national authorities.

A long-term multidisciplinary pilot research in a natural habitat of wild emmer wheat, *Triticum dicoccoides* (Koern.) Aarons. [*T. turgidum* L. var. *dicoccoides* (Koern.) Bowden] has been initiated in Israel, starting in 1984 with the aim to develop an applicable methodology of *in situ* conservation. In the chosen site (Ammiad - northern Israel) studies have been carried out along four transects totalling about 800 m. The main scientific questions approached in this study are summarized by Noy-Meir *et al.* (1989). Results obtained in the first three years are detailed in Anikster (1988) and a series of summaries for six years of research, up to the present, are now in press (Anikster and Noy-Meir, 1991, Anikster *et al.*, 1991, Felsenburg *et al.*, 1991 and others). *In situ* conservation in general and the wild emmer project in particular were discussed in an international workshop on *in situ* conservation (cf. Israel Ministry of Science and Technology, 1990).

The multidisciplinary research comprises a detailed ecological characterization of the study site in general and of
the numerous microsites. The genetics of 250 lines, sampled
along the transects and grown from seeds, have been studied by
various methods, including allozyme investigation. Morphological and phenological characters, as well as some of those
connected with the yield (e.g., seed proteins), have been
quantified. Resistance to soil-borne and foliar diseases have
been compared between lines. The dynamics of the population
has been assessed by comparison of plants collected each year
from near to the exact locality from which the seeds for each
line had originated.

The amount of information obtained, though quite considerable, is based on a single species in one habitat. In order
to be able to propose a methodology for the conservation of
wild emmer wheat and consequently for species similar in some
important features (like breeding system) and later for being
able to form general guidelines for *in situ* conservation, an
extension of the study in various directions has been proposed
in the above mentioned workshop. Studies, similar to those
made in wild emmer are needed in several habitats, in particular in habitats with extreme ecological characters. Areas for
in situ conservation should be chosen to include disease-
favorable and disease-unfavorable sites for foliar and air-
borne diseases. In order to devise some more general guidelines, one has to compare population studies in several
species with breeding systems (obligatory outbreeders, obligatory selfers) that differ from that of the wild emmer, which
is mainly a selfer. These studies should be applied to other
families in which genetic resources have contributed in the
past, or may contribute in the future, to cultivation (e.g.,
legumes). Management of reserves in which *in situ* conservation of populations are maintained are of great importance.
However, management practices can be proposed only after testing different regimes in various habitats.

It is obvious that the resources needed for scientific
manpower on a long-term basis for studies leading towards a
methodology for *in situ* conservation are very large. Furthermore, the cost of the land in areas to be conserved may make,

at least in some parts of the world, such conservation practice even more expensive. One could perhaps economize in all
respects by using, at least for pilot studies, existing nature
reserves, with preference for those sites that include populations of several species with genetic resources of interest to
mankind.

Although some of the agencies that support the wild emmer
project of Israel (Committee on Managing Global Genetic
Resources, 1991) favor extending the project, it seems that
because of the investment still needed, *in situ* conservation
has at present only very slight immediate chances to be
adopted as a standard procedure for conservation of genetic
resources. There is no doubt about the investment being
worthwhile in proportion to the expected advantage because *in
situ* conservation preserves a wide genetic diversity within a
dynamic system which cannot be achieved by any other method.

We would like to propose some ways for optimal use of *in
situ* conservation, as it parallels efforts to reduce the loss
of existing genetic resources:

1. The methodology for *in situ* conservation should be
established by several additional pilot studies (including
those needed for reaching a full evaluation of the wild emmer
wheat research);

2. Priorities should be set up for those species and populations for which dynamic conservation is most needed, screening the available genetic resources of relatives of cultivated
plants and of plants with potential economic value;

3. Habitats of endangered species could be conserved first
and studied later. In fact, many plants already screened for
possible use in the future, are in regions where the limitation of land may be only a weak constriction - so first of all
several of their habitats could perhaps be reserved, even as a
temporary measure, up to the time when the value of the tested
plants will be better known and they can be studied for either
in situ or *ex situ* conservation;

4. Studies for *in situ* conservation, by being mainly concerned with understanding the genetic diversity of investigated species and populations, could provide also an optimal
model for *ex situ* conservation of the germplasm.

Literature Cited

Annikster, Y. 1988. The biological structure of native populations of wild emmer wheat (*Triticum turgidum* var. *dicoccoides*) in Israel. Final Report to USDA (ARS), Oregon
State University and National Council for Research and
Development, Israel.

_____ and I. Noy-Meir. 1991. The wild wheat field laboratory
at Ammiad. *Israel J. Bot.* (in press).

_____, A. Eshel, S. Ezrati, M. Feldman and A. Horovitz. 1991.
Patterns of phenotypic variation in wild tetraploid wheat
at Ammiad. *Israel J. Bot.* (in press).

Committee on Managing Global Genetic Resources. 1991. *The U.S
National Plant Germplasm System*. National Academy Press,
Washington D. C.

Felsenburg, T., A. Levy, G. Galili and M. Feldman. 1991. Polymorphism of high-molecular weight glutenins in wild tetraploid wheat in a native site. *Israel J. Bot.* (in press).

Israel Ministry of Science and Technology. 1990. *International
Workshop: Dynamic In Situ Conservation of Wild Relatives
of Cultivated Plants*. (Scientific Program and Abstracts).

Noy-Meir, I., Y. Anikster, M. Waldman and A. Ashri. 1989. Population dynamics research for *in situ* conservation: wild
wheat in Israel. *FAO/IBPGR Plant Genetic Resources Newsletter* 75/76: 9-11.

Zohary, D. 1983. Wild genetic resources of crops in Israel.
Israel J. Bot. 32: 97-127.

SCREENING VALUABLE GENES FROM WILD SPECIES OF PLANTS

Lin Zhong-ping

Institute of Botany, Academia Sinica, Beijing 100044, China

Hu Zhong
Kunming Institute of Botany, Academia Sinica
Kunming 650204, China

and

Ye He-chun
Institute of Botany, Academia Sinica, Beijing 100044, China

Summary - Three research projects are reviewed which show
the utility of screening for valuable genes in wild species.
The projects are: cell lines of *Arnebia euchroma* have been
established that yield over 14% (dry wt.) of shikonin. Additional shikonin may be obtained by modifying the biosynthesis
pathway by recombinant DNA technology. Protein anti-fungal
protein (PAFP) has been isolated and characterized from *Phytolacca americana*. The PAFP gene may be used in genetic engineering to improve crop fungal resistance. A genomic library
of *Glycine soja* is being constructed with the goal of enhancing the quality of protein in soybean. Wild relatives of
soybean, native to China, are being screened for protein and
these strains will be used in the program.

INTRODUCTION

It is important to note that during the civilized history
of human beings, man introduced wild plants into cultivation
and obtained great benefits from them. However, to date, only
a small portion of the estimated 300,000 species of plants has
been utilized by man, with a considerable number of species
still unknown to man. Thus, the natural wealth is abundant in
the wild plant resources. Wild plants not only provide us
with the amenity of environment but a great deal of useful
substances. Unfortunately, many valuable wild species have
been made extinct by human activities (Wang and Chen, 1990).

What can we do for the conservation and utilization of wild plant resources at the molecular level? Transferring of foreign genes into recipient plants, (see workshop proceedings) to produce genetically modified transgenic plants will become a popular technique in plant breeding in the next century.

The exploitation of genes from wild species will call attention to conservation and protection of wild species. In this respect encouraging progresses have been recently made. For example, the gene encoding sulphur-rich protein has been isolated from Brazilian nut for the purpose on improving the nutritional balance in legumes (Altenbach et al., 1989). Another example is the very sweet protein, thaumatin, extracted from the arils of fruit of *Thaumatococus danielli* which grows in West Africa. Thaumatin genes have been cloned and will be transferred into some other crops to get sweet food with low caloric value (Edens et al., 1982).

CURRENT RESEARCH

In order to search and screen for valuable genes from wild plants, three research projects on different species are currently in progress in our laboratory.

1. *Arnebia euchroma* - *Arnebia euchroma* (Boraginaceae) is distributed over the Xinjiang and Tibet provinces of China, but not very common in nature. No success has been obtained to introduce it into cultivation yet. *Arnebia euchroma* has been used as a traditional Chinese medicine for burns, wounds, frostbite, skin ulcers and anal hemorrhage (Heide and Tabata, 1987). Recently, anticancer activity has been found in shikonin and its derivative, acetylshikonin. *Arnebia euchroma* has now been listed as one of the protected plants by the Chinese government. Because of the exhaustion of wild plants and difficulty of domestication, a large scale cell culture of it is being carried out, which can produce the effective secondary substances. One cell line of *Arnebia euchroma* has been selected that produces shikonin derivative 7-8 times higher than in roots. The total contents of shikonin derivatives are 14.26% (dry weight) in the fermentation culture. The biosynthetic pathway leading to shikonin and various physical and

chemical factors influencing the production of shikonin have been investigated (Ye and Li, 1991). It has been defined that phenylalanine ammonia lyase (PAL) and p-hydroxybenzoic acid geranyltransferase (PHB geranyltransferase) are two key enzymes in shikonin biosynthesis. We are proceeding to clone the gene encoding for PHB geranyltransferase. It would be of great potential in promoting shikonin biosynthesis by means of biotechniques. A chimeric plasmid containing the PHB geranyl-transferase (or PAL) gene under control of CaMV 35S promoter will be constructed and transferred into the cells of *Arnebia euchroma*.

2. *Phytolacca americana* (Pokeweed) - Several proteins of physiological activities have been reported to be isolated from *Phytolacca americana* (pokeweed). Pokeweed mitogen (PWM) from roots is used as a mitogenic drug for both T and B lymphocytes. Pokeweed antiviral protein (PAP) from leaves and seeds are ribosome inactivating protein (RIP) (Barbieri *et al*., 1982). We have isolated and purified a protein named PAFP-s with antifungal activity from the seeds.

PAFP-s has the molecular weight of 7 Kd and is rich in Gly, Cys, Arg and Tyr, but lack of Trp, Met, Thr, His and Leu. The amino acid sequence of PAFP-s is:

```
H-A G C I K N G G R C N A S A G
  P P Y C C S S Y C F Q I A G Q S
  G V C P N R R G R C Y Q R D Q E
  G P Y P G P V Y K C Q C F G F R-OH
```

PAFP-s causes a visible inhibition of the growth of *Trichoderma viride* at the concentration of 1 μM of the protein in medium. PAFP-s also inhibits the growth of a mushroom *Morchella conica* Pers., *Fusarium oxysporum* f. *vasinfectum*, a pathogen of cotton, and *Pyricularia oryzae*, a pathogen of rice at 10 μM. PAFP-s is thermo-stable. The antifungal activity retains after heat treatment at 95°C for 30 minutes (Hu *et al*., 1990).

PAFP-s gene might be used in the genetic engineering of improving crop tolerance to fungal diseases. For the purpose of isolation of PAFP-s gene, two DNA fragments have been synthesized according to the plant codon usage pattern. Random

primers using these fragments have been used as the probes to
screen the coding sequence of PAFP-s from the *Phytolacca americana* genomic library. Some of positive clones containing
Phytolacca DNA fragments homologue to the probes have been
obtained. The DNA sequencing is already in progress.

3. *Glycine soja* and *Glycine gracilis* - Wild soybean is
native to China. The distributions of wild species extend
over the Northeastern provinces and along the valleys of the
Yellow River and the Huaihe River. They belong to *Glycine
soja* and *G. gracilis* and the latter is closer to the cultivated species *G. max*. One of my co-workers used a strain of
wild soja as a paternal plant to cross with the cultivated
soja and eventually obtained many strains with high contents
of proteins in the progenies. In addition, he used the total
DNA of the selected strains of wild soja together with the
pollen of cultivated soja to pollinate the stigmas and produce
strains of soja with the characteristics of wild types,
including the characteristic of high protein content. Eight
strains of soja have already been grown to the fifth generation and no segregation occurred. Their protein contents are
higher than 45%, compared to the control lines that contain
between 38-40% protein (Lei *et al.*, 1989). We have constructed a genomic library of *G. soja* and screened out a
genomic clone of conglycinin α'-subunit (Zou *et al.*, 1989).
Now we have found that the storage protein genes of the Leguminosae are genetically conservative and stable. There are no
obvious difference between the cultivated and the wild
species. Hereafter, our aim is to isolate the species - specific DNA sequence of *G. soja*. There are numerous reports of
repeat DNA sequences specific to certain genera and species.
These repeat sequences are of significant importance in evolution (Zhao *et al.*, 1989; Hallden *et al.*, 1987). For example,
some species of *Actinidia* possess specific repeat DNA
sequences. Dr. R. C. Gardner of Auckland University, New Zealand has isolated a genome-specific repeat sequence from *Actinidia deliciosa* var. *deliciosa* (Crowhurst and Gardner, 1991).
Chinese scientists have already collected and cultivated several thousand taxa of wild soja and are analyzing their
genetic diversities (Lin *et al*, 1986, 1987; Hu and Wang,

1985), which is fundamental for further research work at the
DNA level.

We hope that in the aspects of protection and utilization
of gene resources of wild plants, we can cooperate more
closely with the colleagues from other countries.

Literature Cited

Altenbach, S. B., K. W. Pearson, G. Meeker, L. C. Staraci and
    S. S. M. Sun. 1989. Enhancement of the methionine content
    of seed proteins by the expression of a chimeric gene
    encoding a methionine rich protein in transgenic plants.
    Plant Mol. Biol. 13: 513-522.
Barbieri, L., G. M. Aron, J. D. Irvin and F. Stirpe. 1982.
    Purification and partial characterization of another form
    of antiviral protein from the seeds of *Phytolacca ameri-
    cana L.* (Pokeweed). *Biochem. J.* 203: 55-59.
Crowhurst, R. N. and R. C. Gardner. 1991. A genomic specific
    repeat sequence from kiwifruit (*Actinidia deliciosa* var.
    deliciosa). *Theor. Appl. Genet.*. 81: 71-78.
Edens, L., L. Heslinga, R. Kolk, A. M. Ledeboer, J. Maat, M.
    Y. Toonen, C. Visser and C. T. Verrips. 1982. Cloning of
    cDNA encoding the sweet-tasting protein thaumatin and its
    expression in *Escherichia coli. Gene. 18: 1-12.*
Hallden, C., T. Bryngellson, T. Sall and M. Gustafsson. 1987.
    Distribution and evolution of a tandemly repeated DNA
    sequence in the family Brassiceae. *J. Mol. Evol.* 25:
    318-323.
Heide, L. and M. Tabata. 1987. Enzyme activities in cell-free
    extracts of shikonin-producing *Lithospermum erythrorhizon*
    cell suspension cultures. *Phytochemistry* 26: 1645-1650.
Hu, Z. A. and H. X. Wang. 1985. The biochemical genetics of
    soy-bean. *Soybean Science* 4: 245-248.
_____, X. Z. Liu, Q. Z. Huang, X. M. Zhu and D. Yang. 1990.
    Isolation, characterization and amino acid sequencing of
    an antifungal protein from the seeds of *Phytolacca ameri-
    cana.* *In: Proceeding of China Symposium on Polypeptides*
    (in press).
Lei, B. J., G. C. Yin and Z. P. Lin. 1989. Variation occurring

by introduction of wild soybean DNA into cultivar soybean. *Oil Crops of China 3: 11-14.*

Lin, Z. P., G. C. Yin and B. J. Lei. 1986. Composition of globulins and its relationship to total seed proteins in Glycine. *Science Bulletin 31: 118-122.*

_____, G. Z. Peng and G. C. Yin. 1987. Analysis of amino acid component variation of soybean protein. *Soybean Science* 15: 105-111.

Wang, S. and L. Z. Chen. 1990. Future economic development and biological diversity conservation, sustainable utilization and research in China. Pp. 6-11. In: S. Wang and X. H. Du (eds.), *Proceeding of the Symposium on Biological Diversity*, Bureau of Biosciences and Biotechnology, Chinese Academy of Sciences, Beijing.

Ye, H. C. and G. F. Li. 1991. Studies on the production of shikonin derivatives by using large scale cell culture of *Arnebia euchroma. Botanical Research* (in press).

Zhao, X, T. Wu, Y. Xie and R. Wu. 1989. Genome-specific repetitive sequences in genus *Oryza. Theor. Appl. Genet.* 78: 201-209.

Zou, J. T., Z. P. Lin, Y. C. Yin and Q. Y. Qian. 1989. Preparation of large fragment chromosomal DNA and construction of genomic library from *Glycine soja. Chinese J. Bot.* 1: 161-163.

THE STATUS OF RARE AND ENDANGERED PLANTS IN CHINA AND EFFORTS FOR THEIR PROTECTION

Ma Cheng

Institute of Developmental Biology, Academia Sinica
Beijing, China

Lin Zhong-Ping and Hong De-Yuan

Institute of Botany, Academia Sinica
Beijing, China

Summary - The status of the flora of China is reviewed and the ratio of deforestation and plant extinction is shown to be greater than the world average. Six specific proposals are presented for saving the living organisms of China.

INTRODUCTION

The human species appeared on the earth, at a period of a very high level of biological diversity. Today, at a time of population expansion and transformation of natural environment, biological diversity has been decreasing to the lowest level since the Mesozoic period. Certainly, the result of this kind of biological conflict is very serious. This crisis of biological diversity actually may imperil the existence of human beings. In recent years, this questions has attracted the attention all over the world, as well as in China.

THE IMPORTANT POSITION OF CHINESE PLANTS

The climate in China falls into tropical, subtropical and cold-temperate zones; the diversity of these climatic zones is unique in the world. Topographically, extending from the eastern seaboard westward to the roof of the world-the Qinhai-Xizhang plateau, from ground level to the altitude of 8884 m, the complexity of its topography is also unique in the world. The diversity of the climate and the complexity of the topography have created optimal conditions for the growth of various plants. *The Illustrated Handbook of Chinese Higher Plants*

(Anonymous, 1972-75) includes 13,000 species of plants, all of which have economic value or scientific significance. There are approximately 80 volumes planned for the *Flora of China* (Acad. Sin. Ed., 1959-90), of which 58 volumes had been published. Work is progressing on publishing additional volumes. From these, we can estimate that there are about 30,000 species of higher plants in China. In the world, due to the large areas of subtropical rain forests in Amazon, Brazil is the champion in the number of species of plants. The number of plant species in Peru and Indonesia is about the same as in China. However, China ranks first in the Northern Hemisphere in the diversity of plant species.

The plants in China not only are abundant in species diversity, but also are at a very important position. As we know, China is one of the origins of cultivated plants in the world. Thus, China has very rich germplasm resources with large quantities of wild-parent species of cultivated plants, such as soybean, rice, barley, tea, etc. For example, in just one area, Xishuangbanna, there are about 100 wild-parent species of cultivated plants These constitute a precious botanical wealth for China. Ten years ago, a variety of wild soybean with dense, white hair was introduced into the United States, where it was hybridized with cultivated soybeans. The new variety has now been developed with drought-resistance and can grow on poor, dry soil, saving 15% more water than the cultivated soybean.

China is also a great treasure-house of medicinal plant resources with nearly 7000 species. In recent years, cancer, AIDS, and cardiovascular diseases are becoming more and more serious in damaging the health of human beings. Every country in the world is paying more attention in selecting the best medicines from wild plants for the treatment of cancer and cardiovascular diseases. Some institutions of the Chinese Academy of Sciences, such as the Shanghai Institute of Materia Medica, Kunming Institute of Botany, Institute of Botany in Beijing and Chengdu Institute of Biology are working on the selection of effective components with the cooperation of institutions in the United States and Japan. Last year, scientists in San Francisco found that the Chinese medicine *Tri-*

chosanthes kirilowii has been effective in killing AIDS' virus without toxicity to healthy cells. There are 40 types of *Trichosanthes kirilowii* in China. To study, develop, utilize and protect these plants will be very important research projects.

China is also a kingdom of flowers. Some famous flowers, such as rhododendron, camellia, *Paeonia suffruticosa*, *Camellia chrysantha*, *Primula* and *Lilium* are originally from China. Especially when one is on an expedition in the southwestern regions of China, either walking in the forest of camellia at low altitude, bushes of rhododendron at middle altitude, or climbing up the alpine meadow about 4000 m altitude, one will see flowers of *Primula*, *Pedicularis* and *Meconopsis* all over the mountains and plains, as if one is living in the sea of flowers. There are nearly 500 species of rhododendron, more than 300 species of *Primula* and over 100 species of camellia in China. An American horticulturist, E.H. Wilson, once said that the Chinese flowers are the mother of the gardens of the world, and the Europeans said that without Chinese flowers, there will be no gardens in the world (Hong *et al.*, 1990).

In addition, many ancient and old plants, called living fossils, are still growing in China. The famous living fossil *Gingko biloba* appeared as early as the Jurassic period. It used to be widely distributed around the Northern Hemisphere. Now, the worldwide cultivated plants of *Ginkgo biloba* are generally considered to be of Chinese origin. It has been reported that there are wild *Ginkgo biloba* growing in Mt. Tienmon in Zhijiang and mountain areas of the middle parts of Hubei. Some disputes are still going on among the botanists. *Abies* used to be widely distributed around the Northern Hemisphere, and was thought that it had all become fossilized. However, two botanists Prof. Hu Xianxiao and Prof. Zheng Wanjun have found that *Abies* still grows at Modaoqu between the border of Hupei and Sichuan. This discovery has created a great sensation throughout the botanists of the world. *Abies* had been introduced in almost every botanical garden in the world in 1948. *Cathaya argyrophylla* is also a famous living fossil. It was found by Profs. Chun Huanyong and Kuang Keren in 1958 (Chun and Kuang, 1958), and at the present time there is not a specimen in the herbaria abroad. *Cathaya argyroph-*

ylla is occasionally now found in Sichuan, Guichou, Hunana and
Guangxi. All this shows the great scientific value of rare
and endangered plants in China. The Chinese plant resources
hold a place that cannot be neglected in the world.

THE RARE AND ENDANGERED STATUS OF CHINESE PLANTS

Since the appearance of life on the earth, large scale
biological extinction has occurred several times. The main
reason for this is due to the competition for existence and
the change of geological environment. Every sign shows that
at present, living things on the earth are facing another new
and large-scale extinct period through the intervention of
mankind beings.

As Chinese plants have such a special position, their
existence is still facing serious crisis. Over 30 years,
because of the expansion of the population, the development of
industry, lack of control, over-exploitation and utilization,
the speed of destruction of forests has been very quick. During the early 1950's, Hainan Island had 25.7% of areas covered
with forest; now only 7.2% is left. 70% of the forest areas
has disappeared, decreasing progressively 2.7% every year. In
Xishuangbanna, 55% of the land was covered with original forests during the 1950's; now only 28% exists. The status in
other regions is even worse. The mountain regions in the
southern part of China, especially in the southwestern part,
used to be covered by green hills. Original forests and well-
developed natural forests also occurred in western Hubei,
western Hunana, Guichuo, western Guangxi, Yunnan and western
Sichuan. Now most of the forestry bureaus can either work
only on the top of the mountains, or have nothing to cut. The
farmers on the mountains had been using slash-and-burn cultivation and cut small trees for cooking. As a result, most of
the mountains have become barren. At present, we have approximately 400 natural conservation districts in China which
occupied 2% of the total area in China. Only a few of the
original forests and fairly developed natural forests remain.
Within several decades, the speed of forest destruction in
China has been greater than the world average of 1% per year.
The exact rate is not known. Using a rate 2.7% of forest area

loss every year in Hainan Island to estimate the forest decreasing percentage of the whole country, the rate is surely not lower than 2.7% per year. Between the end of the 1950's and early 1960's, the Yangtzi River was still a green-colored river; only 30 years later, it had turned into China's second yellow river. During the 1980's tremendous floods occurred twice on the upper and middle reaches of the Yangtzi River. From these events we can see how rapid is the destruction of forests in China.

Due to the change in the environment, the original ecological system has been destroyed, and living things are threatened and endangered. The book *Rare and Endangered Plants in Guang Dong, China*, (Wu, 1988), listed 105 species of rare and endangered plants, composed mostly of tropical and subtropical rain forest plants. In the book *Rare and Endangered Plants in Guizhou, China* (Huang, 1989), the species described were composed of evergreen broad leaf forests. The survey by the biologists of Shanxi Province revealed that 44 species of rare and endangered plants distributed in Shanxi are also facing extinction.

In addition to plants, animals are also facing the same fate. The Asiatic two-horned rhinoceros used to be distributed in the southern part of Yunnan, but in the 1930's, the last one was killed. As a result, the Asiatic two-horned rhinoceros is now extinct in China. Para was found in western Yunnan in the Gos, but now it is also extinct. The survey of the Kunming Institute of Zoology in 1983 revealed there had been 97 species of birds in the forest of Xishuangbanna, but now there are only 28 species, a decrease of 71% within ten years.

According to the general world estimate, about 10% of the world's species are endangered. As mentioned above, the speed of the environmental damage in China is above the average world rate. In China the denudation of plant resources is very serious. We can be sure that the percentage of endangered plants is greater than 10% and might be as high as 15-20%. According to this estimation, the endangered higher plants will number 4000-5000 varieties. Looking from another point of view, the list of endangered plants that we are collecting already contains approximately 3000 species, with the listing still not complete. Certainly this number is only a rough estimate. The very urgent task before us at present is to make a general survey of the endangered plants, get their exact names, study their biological characters and draw up scientific methods to protect them.

Not only are 4000-5000 plants in China endangered or threatened, a large number of plants have already become extinct. The precise number may never be known. With the investigation of several endangered plants, we have found that some plants, such as *Ormosia howii*, *Hopea mollissima*, *Nyssa yunnanensis*, *Carallia diplopetela*, *Dunnia sinensis*, *Mussaenda anomata*, *Diplandrorchis sinica* and *Archinecttia gaudissartii* still cannot be found though we had investigated for several years. Most of the species have only a single individual left or exist as several individual plants, such as *Carpinus putoensis*, *Ostrya rehderaina*, *Abies beshazuensis* in Zhejiang and *Bhesa sinenis*, *Apterosperma oblata*, *Furyodendron exelsum* and *Chuniophoenix hainanensis* on Hainan Island. Based on preliminary statistics, about 5% of the plants listed in the rare and threatened plants have gone extinct during the past few decades. Deducing from this estimation, it appears that at least 200 species of plants have recently become extinct.

The results of statistics from the biological community revealed that one plant species coexists with 10-30 other living beings by providing their food and environment. The extinction of one plant species will threaten the existence of 10-30 other living things. According to this reasoning of calculation, from the 4000-5000 species of plants that are on the endangered list in China, it is estimated that 40,000 - 50,000 species of other living beings will be threatened. Given the estimation above that 200 species of plants were already extinct, it is possible that 2000-6000 other living beings no longer exist. Facing this situation, it would be a crime not to adopt any effective measures to rescue these endangered plants.

THE SUGGESTIONS FOR SAVING LIVING BEINGS

In order to improve the conditions for plant conservation

in China, we suggest that the following measures be adopted:
(Chen, 1989; Xu, 1987):

1) Although China has a network of protected areas, a more
comprehensive network needs to be established. The main problems, at this time, are how to manage the already established
nature reserves, how to implement the training of their admin-
istrators and how to enable them to become centers or laboratories of nature in a coordinated and concerted effort. These
problems should be resolved quickly.

2) The introduction of threatened plants into botanical
gardens is a strategic measure to protect them from extinction. China has just begun cultivating some rare and endangered plants in a few botanical gardens, but as a whole, it is
not well organized. The next step if to establish a center
for plant conservation and to organize a network of all the
botanical gardens in China. Efforts should be made to introduce more plant species into local botanical gardens and then
reintroduce them into the wild.

3) It is the greatest urgency to strictly control wildlife
and plant trade, and to strengthen public education. Further
emphasis should be placed on raising the awareness of the public, including public officials for the economic, ecological
and scientific significance of wildlife conservation for mankind.

4) In addition to the loss of forests and the destruction
of their habitats, the occurrence and distribution of plants
are caused by their internal factors. We need to ascertain
the status of each species and learn why it has become rare or
endangered, to understand the many facets of its biology and
then to take measures to protect it. The compilation of the
China Red Data Book is merely the first step. Most of the
taxa included are in urgent need of further study.

5) There is a need to establish a DNA Bank for rare and
endangered species. Many plant have very important values in
economy and science. It would be a great pity if these plants
were lost or on the verge of extinction before they could be
thoroughly studied and exploited. In order to study, save and
make the extinct biological resources reappear on the earth,
we must take measures to save their genetic substance by the

storage of DNA and cells. With the encouragement from DNA
Bank-Net (R. P. Adams, pers. comm.), we have started the
establishment of a bank for rare and endangered plants. This
is a very important and significant research project.

6) There is a need to establish a plant cell bank. The
techniques to regenerate a plant by using cells, tissues and
organ culture is no longer a problem for us. Experiments on
the preservation of plant cells, tissues and organs are now
being carried on with the establishment of a cell bank in the
Institute of Botany, the Chinese Academy of Sciences. Preliminary results have been obtained by storing the plants at
ultralow temperature to stop all metabolic activities and then
reviving the plants.

Large scale set of animal cell bank have been established
in the Shanghai Institute of Cell Biology and Kunming Institute of Animals of the Chinese Academy of Sciences. Due to
the variety of animal species in southwestern China, a wild-
animal cell bank has been established in the Kunming Institute
of Animals for the purpose of the conservation of animal
genetic resources and their later utilization. The animal
cell bank now is known as a frozen animal garden. At present, 170 species of animals have been saved in this animal
bank, included cells ranging from insects to human beings,
among which are the cells of special rare and endangered
species such as the golden monkey, black muntjae, tufted deer,
red goral, mouse-deer, etc.

With the advance of cell biology and developmental
biology, the secret of cell differentiation and individual
development will be finally revealed some day. Through the
techniques of cell culture and nuclear transplantation from
DNA and cell bank, people in the future may re-construct the
plants and animals which have become extinct on the earth.

Literature cited

Anonymous. 1972-1975. *Iconographia Cormophytorum Sinicorum*
(5 Vols.). Science Press, Beijing.
Chen, Sing-chi, 1989. Status of the conservation of rare and
endangered plants in China. *Cathaya* 1: 166-180.

Chun, H.-Y and K.-R Kuang. 1958. A new genus of pinaceae, Cathaya Chun et Kuang, from southern and western China. *Bot. Zuln. USSR 43(4): 461-470.*

Del. Fl. Reip. Pop. Sin. Agend. Acad. Sin. Ed. 1959-1990. *Fl. Reip. Pop. Sin.* (58 Vols.). Science Press, Beijing.

Huang, Wei-lian (ed.). 1989. *Rare and endangered plants in Guizhou* (in Chinese). China Environ. Sci. Press, Beijing.

Hong, De-yuan *et al.* 1990. The important position of the Chinese plants and their endangered status (in Chinese). In: Proceedings of the Academia Sinica Symposium on Biological Diversity, Beijing (March 28-31, 1990).

Xu, Zaifu, 1987. *The work of Xishuangbanna Tropical Botanical Garden in conserving the threatened plants of the Yunnan Tropics, Botanic Gardens and the World Conservation Strategy.* Pp. 239-253. Academic Press Inc. Ltd., London.

Wu, De-lin (ed.). 1988. *Rare and endangered plants in Guangdong.* Guangdong Sic. Techn. Press, Guanzhou.

TRANSPORTATION OF PLANT MATERIALS
OUT OF DEVELOPING COUNTRIES

Daniel Abbiw

Botany Department
University of Ghana
Legon, Ghana

Summary - This paper enumerates: the twelve main groups of
plants or plant materials (both raw and processed) that may be
transported out of developing countries and the part or parts
that may be taken; the sources (wild, tended, semi-cultivated
and cultivated) where these materials are normally obtained;
the main factors that are known to contribute to the gradual
decline and eventual extinction of plant species. It estab-
lishes that there is a direct link between the quantity, the
part or parts involved, sources of supply and subsequent
effects on populations of a species. Based on this linkage,
recommendations are made as to what plant part or parts could
be safely transported without a threat to the species and
should not be transported because this could threaten the
species. Finally, the paper outlines the necessary precau-
tionary measures including education, regulations and legislations, by-laws, heavy court fines for defaulters, etc. for
preventing the indiscriminate exploitation of our forests and
woodlands; managing and controlling the forests for sustained
yield; and ensuring the protection and preservation of plant
species.

INTRODUCTION
In spite of advances in modern technology and man-made
substitutes, mankind still depends on plants for survival, and
on plant materials for food, clothing and shelter almost
everyday. Despite this dependence on plants, man's activities
tend to destroy the natural forest and woodland with its
numerous benefits.

These activities include farming (particularly the traditional slash and burn method, with its associated shifting
cultivation), collection or gathering of fuelwood, commercial
timbering, hydro-electric power generation, exploitation of
mineral resources, road-building, housing, industrial pollution and brush fires. With the ever increasing awareness of
these threats to nature, and world-wide concern in protecting
our resources, environmentalists, naturalists, ecologists,
foresters, policy makers and governments are coordinating
their efforts to find solutions to minimize the effects of
these activities on the vegetation and the environment.

In addition to the aforementioned activities, there are
other processes which account for the gradual degradation of
the natural vegetation. For instance, the large scale screening of medicinal plants for possible new sources of drugs
(Ayensu, 1978) means the transportation of many such plants to
other countries. Similarly, the appreciation of the beauty of
decorative plants and man's desire to live in beautiful environments have resulted in the introduction of many exotics to
alien countries (Abbiw, 1990).

Besides the medicinal and decorative values, there are
several other reasons (commercial, economic, industrial, agricultural, scientific, educational, social, etc.) why plant
materials are transported out of developing countries. Whatever the reasons are, and however justifiable they may be, the
transportation of plant material should not be allowed to continue if it degrades the natural vegetation. A solution to
the problem of transportation of plant materials out of developing countries must be found and enforced without further
delay to save the plant species from undue exploitation, and
mankind from the effects of the resulting deteriorating vegetation and weather conditions.

WHAT MATERIALS ARE TRANSPORTED

The plant materials transported out of developing countries may be divided into twelve main groups (nomenclature
from *Flora of West Tropical Africa* by J. Hutchinson and J. M
Dalziel with revision by R. W. J. Keay and F. N. Hepper,
1954-72):

1. <u>Timber</u> - either whole logs or sawn, also timber products. Irvine (1961) lists about 400 timber trees. The most
threatened are the merchantable or marketable species. Of
these the following are the twenty most preferred from the
west coast of Africa: *Khaya anthotheca* (Welw.) C. DC., *K. ivorensis* A. Chev., *K. grandifoliola* C. DC., *Entandrophragma
angolense* (Welw.) C. DC., *E. cylindricum* Sprague, *E. utile*
(Dawe & Sprague) Sprague, *Pericopsis elata* (Harms) Van Meeu-
wen, *Tieghemella heckelii* Pierre ex Chev., *Triplochiton scleroxylon* K. Schum., *Nauclea diderrichii* (De Wild. & Th. Dur.)
Merrill, *Lovoa trichilioides* Harms, *Heriteira utilis* Sprague,
Milicia (Chlorophora) *excelsa* (Welw.) C. C. Berg, *Mansonia
altissima* A. Chev., *Turraeanthus africanus* (Welw. ex C. DC.)
Pellegr., *Guarea cedrata* (A. Chev.) Pellegr., *G. thompsonii*
Sprague & Hutch., *Aningeria robusta* (A. Chev.) Aubrev. & Pel-
legr., *Canarium schweinfurthii* Engl. and *Mitragyna ciliata*
(DC.) O. Ktze.

2. <u>Charcoal</u> - In Ghana, the export of charcoal to Saudi
Arabia (mostly) had to be suspended because of public outcry
on its impact on the environment. Presently only charcoal
manufactured from hardwoods logged from the degraded Subri
Forest Reserve may be exported.

3. <u>Cash Crops</u> (raw or processed) - The main cash crops
from the west coast of Africa are *Theobroma cacao* Linn.
(cocoa), *Colfea arabica* Linn. (and other species of coffee),
Vitellaria paradoxa Gaertn. F. (Shea butter), *Cocos nucifera*
Linn. (coconut, copra), *Elaeis guineensis* Jacq. (Oil palm),
Gossypium Linn. species (cotton), *Hibiscus cannabinus* Linn.
and *H. sabdariffa* Linn. (bast fibre), *Cola acuminata* (P.
Beauv.) Schott & Endl. and *C. nitida* (Vent.) Schott & Endl.
(Cola nuts), *Nicotiana tabacum* Linn. (tobacco), *Saccharum
officinarum* Linn. (sugar-cane) and *Hevea brasiliensis* (Kunth)
Mull. Arg. (Para rubber).

4. <u>Bamboo and Rattan Palm</u> (processed and unprocessed) -
The bamboo poses no problem since it regenerates or coppices
easily. Representative species in the region are *Bambusa
vulagris* Schrad. ex Wendl. and *Oxytenanthera abyssinica* (A.
Rich.) Munro. However, the rattan palm does present a problem. Dalziel (1937) lists the species of climbing palms as

Calamus deeratus Mann & Wendl., *Laccosperma* (Ancistrophyllum)
secundiflorum, *L. opacum*, *L. laeve*, *Eremospatha hookeri* (Mann
& Wendl.) Wendl., *E. wendlandiana* Dammer ex Becc. and *E.
macrocarpa* (Mann & Wendl.) Wendl. Because climbing palms,
like the palm trees, do not regenerate when the stem is cut
and the entire requirements for basketry and allied crafts are
harvested from the wild, all climbing palms are threatened.

    5. Medicinal Plants - Usually the parts transported are
seeds or fruits, leaves, flowers, bark, roots or tubers and
whole plant. The exploitation of the bark, roots, tubers and
the whole plant are a threat to the species if collected from
the wild. Exporting a container full of *Centella asiatica*
(Linn.) Urb. (Indian Navelwort), or *Aloe buettneri* A. Berger
from the wild for instance would exhaust the entire Ghana population. However, the export of seeds and fruits may not pose
a threat to some species. For instance, about 100 tons of
Griffonia simplicifolia (Vahl ex DC.) Baill seeds are exported
from Ghana annually (Abbiw, 1990). Similarly, about the same
tonnage of *Voacanga africana* Stapf seeds are being exported
annually, mostly from the wild, but the species is not endangered.

    6. Drugs, poisons, dyes, etc. - Again the parts usually
transported are seeds, leaves, flowers, bark, roots and tubers, and whole plants. As above, the use of bark, roots, tubers and the whole plant pose the greatest threat.

    7. Decorative Plants - Examples are orchids, aroids and
members of the *Agavaceae* family. The parts of decorative
plants transported are seeds, cuttings, bulbs, corms, rhizomes
and whole plants. Orchids, such as *Eulophia cristata* Steud
are very decorative. Dalziel (1937) observed that they have a
chain of potato-like tubers from which they can be propagated,
making them easily transportable to Europe. Some orchids are
threatened by this transportation. For instance the exportation of *Plectrelminthus caudatus* (Lindl.) Summerh. and *Ancis-
trocladus rothschildianus* O'Brien, two beautiful epiphytic
orchids, and, to some extent the destruction of their natural
habitat through farming activities, have exterminated these
species (Abbiw, 1990).

    8. Sweeteners - The three main plant sweeteners trans-

ported from the west coast of Africa are *Dioscoreophyllum cumminsii* (Stapf) Diels (Guinea Potato or West Africa Serendipity Berry), *Thaumatococcus daniellii* (Benn.) Benth. (Katemfe) and *Synsepalum dulcificum* (Schum. & Thonn.) Daniell (Miraculous Berry) (Enti, 1975;1979). In all three plants the fruits are utilized.

9. <u>Food</u> - The main food items transported from West Africa are *Dioscorea* Linn. species (yams), *Zea mays* Linn. (maize), *Xanthosoma mafaffa* Schott (cocoyam), *Musa paradisiaca* Linn. (plantain) and *Manihot esculenta* Crantz (cassava - processed into gari and kokonte, both which store well and are conveniently transported). They form the major part of food parcels by air from Ghana to Europe, United Kingdom and the United States) (Abbiw, 1990).

10. <u>Fruits</u> - *Psidium guajava* Linn. (guava), *Mangifera indica* Linn. (mangoes) and *Ananas sativus* Schult. F. (pineapple) are being cultivated for export. Other fruits are *Musa paradisiaca* var. *sapientum* Linn. (bananas) and *Persea americana* Mill. (avocado pear).

11. <u>Spices and Flavorings</u> - Among those of commercial important that are transported include *Myristica fragrans* Houtt. (nutmeg), *Aframomum melegueta* (Roscoe) K. Schum. (Guinea grains or Melegueta), *Zingiber officinale* Roscoe (ginger), *Piper nigrum* Linn. (black pepper) and *Xylopia aethiopica* (Dunal) A. Rich. (Ethiopian pepper). Other species are *Monodora myristica* (Gaertn.) Dunal (Calabash nutmeg), *Cymbopogon nardus* (Linn.) Rendle (Citronella), *C. citratus* (DC.) Stapf (Lemon grass) (both processed), *Cinnamomum zeylanicum* Nees (cinnamon) and *Ocimum canum* Sims (American basil). Commenting on the history of Melegueta exports, Lock *et al.* (1977) observed that most of the material exported went to the United Kingdom or other European countries, with some going to the United States.

12. <u>Other Products</u> - Gum acacia, obtained from *Acacia senegal* (Linn.) Willd., *A. dudgeoni* Craib ex Holl. and other species in the genus; resin from *Allanblackia parviflora* Oliv., *Amphimas pterocarpoides* Harms and *Canarium schweinfurthii* Engl. (Incense tree, among others); and gum copal from *Balanites wilsoniana* Dawe & Sprague, *Daniellia ogea* (Harms)

Rolfe ex Holl. (Gum Copal Tree) and *D. oliveri* (Rolfe) Hutch.
& Salz. (African Copaiba Balsam Tree) among several others,
may be transported. Gum copal is one of the first exports of
Ghana with 500 tons in 1876 (Abbiw, 1990). Gutta-percha is
obtained from *Ficus platyphylla* Del. (Gutta-Percha Tree), *F.
sycomorus* Linn. and *Chrysophyllum perpulchrum* Nildbr. ex
Hutch. & Dalz and is exported.

The transportation of these plant materials has been going
on ever since trade routes were established between the old
world and what was then little known or unknown parts of the
world; but particularly so during the industrial age, and more
particularly so after the Second World War.

SOURCES OF THE PLANT MATERIALS

Plant materials are either obtained from the wild or they
are cultivated or semi-cultivated. Some wildings may be
tended. For example, exported hardwoods and *Vitellaris paradoxa* Gaertn. F. (shea butter) are obtained from the wild;
Elaeis guineensis Jacq. (oil palm) is cultivated, while *Catharanthus roseus* (Linn.) G. Don (Madagascar Periwinkle), important in medicine for the treatment of leukemia, is often semi-
cultivated or tended. Burkill (1985) observes that clinical
interest in the plant dates to the 1950's when certain alkaloids were isolated that hopefully would provide control of
Hodgkin's disease and chorio-carcinoma and carcinoma of the
breast.

THE THREAT

The threat to plant species that are transported depends
on the following factors:
1. the sources from which the materials are obtained
2. the part or parts of the plant required.

Plant materials obtained from the wild are the most
threatened if steps are not taken to replace them; unfortunately, this is the present situation. Furthermore, the
removal of whole plants, bark, roots, tubers, corms, rhizomes,
etc. poses more threat to the species than collecting leaves
and fruit. For instance, the over-exploitation of the bark of
Pausinystalia johimbe (K. Schum.) Pierre & Beille, a forest

tree in southern Nigeria for the active yohimbine, an aphrodi-
siac, has virtually wiped out the species in this country.

On the contrary, cultivated plant materials or the transportation of seeds, fruits and leaves generally poses no
threat to a species. As an example, *Theobroma cacao* Linn.
(cocoa beans) have been exported from the west coast of Africa
since the end of the last century, but nevertheless it is
still a leading cash crop in Cote d'Ivoire, Ghana and Nigeria.

It follows that digging out the roots or whole plant of a
quantity of wild plants, or debarking the plant, poses more
threat to the species than the collection of seeds, fruits or
leaves.

FACTORS OF EXTINCTION

The main factors that are recognized as contributing to
the extinction of plant species are the following:

1. Destroying the natural habitat of the species.

2. The continuous removal of plant species without taking
the necessary silvicultural procedures to replant these.

3. The case of frightening and driving away animals of
dispersal or of germination like the elephant, either by killing them or destroying their natural habitat. *Detarium macro-
carpum* Harms, a savanna tree and *Desplatzia subericarpa* Bacq.,
an understory of the high forest, are two such plants that
have been observed to germinate easily in elephant dung, but
not so easily on their own.

4. Traditional beliefs and practices or taboos that sur-
round some plants, and the fear of the consequences of violat-
ing these taboos. An example is the case of *Okoubaka aubrev-
illei* Pellegr. & Normand, a rare forest tree found in Cote
d'Ivoire and Ghana, which is traditionally associated with a
string of taboos (Abbiw, in prep). This tree has the reputa-
tion of killing the surrounding trees (Hutchinson and Dalziel,
1927-36). The generic name among the Anyi tribe of Cote
d'Ivoire means "tree of death". Yet another example is *Ker-
stingiella geocarpa* Harms, an annual legume with edible underground beans. Dalziel (1937) ex. Chev. mentions that in Daho-
mey, it is a food forbidden to women.

MEASURES TO CONTAIN THE SITUATION

Having enumerated the groups of plant materials transported, the part or parts that constitute the most threat to
the species, the sources from where these plant materials may
be obtained and the factors that contribute to the gradual
reduction in the population of plant species or their total
extinction, I would like to outline the necessary measures and
recommendations to arrest plant extinction.

1. Public education - The general public has to be educated on the direct and indirect benefits of the forests and
the need to preserve our forests for the present and future
generations.

2. Recommendations, regulations, legislations and by-laws

a. In Ghana there is already a law banning the exportation of logs from 15 merchantable timber trees. The extension
of this law to cover all the 20 timber trees listed above
would be in the right direction. While these trees account
for about 84% of all exportable timber, there are some 300
other timber trees that are not being exported in quantities
commensurate with their quality and abundance (Abbiw, 1990).

b. In addition to the merchantable timbers, non-
merchantable ones and other plants exploited in commercial
quantities need to be recorded in forest enumeration and stock
survey. Accordingly, their exploitation, like the merchantable timbers, should be based on the sustained yield system.

c. Before any more hardwoods are felled for the manu-
facture of charcoal for export purposes, the tons and tons of
sawdust rotting away year after year could first be compressed
and carbonized into blocks of charcoal to meet our export
needs.

d. All crops (cultivated plants) may be transported
with absolute safety, as these pose no threat to the species.
Plant groups in this category are cash crops, food items,
fruits, spices and flavorings.

e. The seeds, fruits and leaves of wild plants may also
be transported. Plant groups included in this category are
sweeteners and some medicinal plants, poisons, dyes and decorative plants.

f. The transportation of whole plants, bark, roots and

tubers, corms, rhizomes, etc. in commercial quantities from
the wild should be banned by legislation. Any plant exporter
intending to transport these items should first cultivate
them. As a deterrent, defaulters should be made to pay heavy
finds in addition to the confiscation of the plant materials.
Plant groups affected are rattan palm, medicinal plants, poisons, dyes and decorative plants. The export of limited quantities of such plants parts for research or scientific purpose
only may be allowed.

g. As a matter of urgency a special task force should
be set up to list plants that are endangered or threatened,
with the view of collecting seeds, raising up seedlings and
introducing these back into the wild.

CONCLUSION

The transportation of plant materials in commercial quantities from the wild is one of the contributing factors leading to the degradation of the environment, or the decline in
the population of the plants concerned, and their ultimate
extinction as a species.

With public education on the importance of the forests,
and the enforcement of the necessary rules and regulations to
ensure the protection and preservation of our plants, it
should be possible to manage and control the forests for sustained yield to the benefit of the present and future generations.

Literature Cited

Abbiw, D. K. 1990. *Useful Plants of Ghana*. Intermediate Tech-
nology Publications, London and Royal Botanic Gardens,
Kew.
_____. In Preparation. *Traditional Plants of West Africa*.
Aubreville, A. 1959. *La Flore Forestiere de la Cote d'Ivoire*.
3 Volumes. Centre Technique Forestiere Tropical, Nogent-
sur-Marne.
Ayensu, E. S. 1978. *Medicinal Plants of West Africa*. Refer-
ence Publications, Michigan.
Burkill, H. M. 1985. *The Useful Plants of West Tropical*

Africa. Vol. 1 (A-D). Royal Botanic Gardens, Kew.

Dalziel, J. M. 1937. *The Useful Plants of West Tropical Africa.* Vol. 1 (A-D). The Crown Agents, London.

Enti, A. A. 1975. Distribution and Ecology of *Thaumatococcus daniellii* (Benn.) Benth. (Mimeographed).

_____. 1979. Notes on *Synsepalum dulcificum* (The Miraculous Berry). (Mimeographed).

Hutchinson, J. and J. M. Dalziel. 1927-36. *Flora of West Tropical Africa.* 3 Vols. Revision by R. W. J. Keay, Vol. 1 Part 1 (1954), Part II (1958); Vol. II by F. N. Hepper (ed.) (1963); and Vol. III by F. N. Hepper (ed.) Part I (1968); Part II (1972). The Crown Agents, London.

Irvine, F. R. 1961. *Woody Plants of Ghana.* Oxford University Press, Oxford.

Lock, J. M., J. B. Hall and D. K. Abbiw. 1977. The cultivation of Melegueta pepper (*Aframomum melegueta*) in Ghana. *Econ. Bot.* 31(3): 321-29.

STUDIES OF THE RARE AND ENDANGERED PLANT SPECIES IN THE YUNNAN REGION OF CHINA

Zhong Hu and Quan-an Wu

Kunming Institute of Botany, The Academy of Sciences of China
Heilongtan, Kunming, Yunnan, China

Summary - A survey of the diversity of the flora of the Yunnan province is presented. The status of rare and endangered species is reviewed, along with efforts to conserve rare and endangered species in Yunnan.

PLANT DIVERSITY IN YUNNAN REGION

Yunnan is a province located in southwest China. It lies between $21^\circ 8'$ - $29^\circ 15'$ north latitude and $97^\circ 31'$ - $106^\circ 11'$ east longitude with a total of area of 383,000 square kilometers. Because of its complicated geomorphology and the eminent three-dimensional climate differentiation (frigid, temperate and tropical zones), this province is favorable for the growth and development of various kinds of plants. There are 15,000 species, 2136 genera of seed plants and 299 families in Yunnan (Kunming Institute of Botany, 1984) accounting for nearly half of China's total flora. Hence the province is known as the "kingdom of plants". As a rule, in the south and southwest parts of Yunnan (e.g. the Xishuangbanna Autonomous state), there are tropical seasonal rainforests, evergreen monsoon forests and tropical savannah-like vegetation. In the central Yunnan Plateau, the vegetation is mainly drier subtropical evergreen forests. In the southeast, the very special ancient forest types occur on limestone karst. In the northeast, the vegetation is subtropical evergreen forests of the Eastern Asiatic type. The vegetation in northwest Yunnan is often formed by four or five layers of vertical distribution from subtropical pine and oak forests, through the medium temperate mixed broadleaf and coniferous forests, the cold temperate subalpine coniferous forests, to the alpine scrubs, alpine meadows and the ice-bounded vegetation.

The Kunming Institute of Botany, Academia Sinica, is the

center of scientific research studying the exploitation, utilization and protection of the rich plant resources. The
total area of the institute is approximately 1000 hectares.
There are four departments: plant taxonomy and geography, phytochemistry, plant physiology and ethnobotany. The herbarium
of the institute has more than 700,000 sheets of specimens of
seed plants and 120,000 specimens of bryophyte, lichen and
fern. There are two botanic gardens in the institute. The
Xishuangbanna Tropical Botanic Garden, with an area of 900
hectares and living collections of 2500 species of seed
plants, engages in the introduction, acclimatization and conservation of tropical plants. The Kunming Botanic Garden,
with an area 50 hectares, engages in the introduction and cultivation of ornamental plants, special economic woody plants
and medicinal plants. Since 1979 a periodical *Acta Botanica
Yunnanica* has been edited and published by the institute and
distributed to 20 countries.

RARE AND ENDANGERED PLANTS IN YUNNAN

The existence of plant species is threatened by overpopu-
lation, overplanting and the destruction of forests. The
forest-covered land of Yunnan in 1989 was approximately 25% of
the total land area, compared to 50% in 1949. Plants of some
species (e.g. *Ostrya yunnanensis*, *Humulus yunnanensis*, *Staphea
shweliensis*, and *Mecodium lijiangensis*) have not been seen in
recent years. The government and people are being informed of
the urgent importance of plant resource conservation and protection. Priority is given to the plants that are rare and
endangered species. The first list of these plants, announced
by the National Environmental Protection Agency of China in
1984, contained 389 species, of which 155 are in Yunnan. The
second list, announced in 1989, contained 400 species, of
which 105 are in Yunnan. Even though it is not easy to identify what is rare and endangered, four aspects have been considered:

1. Phylogenetically relictual genera - As the glacier
process of the Fourth Period influenced the Yunnan province
only slightly, many ancient plants are maintained in Yunnan.
There are 283 species, 10 family of Gymnospermae in Yunnan,

compared to 800 species or 12 families in all the world,
including relictual genera (e.g. four species of *Cycas*, *Cal-
ocedras*, four species of *Pseudotsuga*, *Torreya*, *Ginkgo*, and
Amentotaxus. Relictual plants of Angiospermae in Yunnan
include the genus Magnoliales, Laurales, Illiciales, Trocho-
dendrales, Cercidphylloles, Eupteleales, Eucoliales, Aristolo-
chiales, Piperares, and Nymphaeiles. There are 15 genera of
the order Magnoliales in the world, 9 in Yunnan; the genera
Manglietiastrumn, *Alicimandra*, *Paramichelia*, and *Tsoongioden-
dro* are distributed in only Yunnan. These plants are significant in the studies of phylogenetics and the evolution of the
plant kingdom.

    2. Special local species - These groups of plants usually
are distributed in limited districts in special micro-
environments and often contain only a single family or genus
and species. They are fragile, but significant in the floristic study of plants.

    3. Tropical flora plants - The tropical seasonal rainfor-
ests of Yunnan are located at the edge of the tropical zone
with a limited latitude and elevation. These types of vegetation are rare in China and have an important role in the ecological system, but are fragile. Ancient families found in
this zone include: Myristicaceae, Dipteriarpaceae, Barringto-
niaceae, Samydaceae, Tetramenaceae, Crypteroniaceae and Son-
neratiaceae.

    4. Economically important species - In the first
announced list of 155 nationally protected species of seed
plants in Yunnan, 100 species were of economic value, including the following: 34 species of woody species (e.g. *Dalbergia
fusca*, *Gmelina arborea*, *Excentrodendron hsienmu*, *Anogeissus
acuminata*, *Nesua ferrea* and *Phoebe zhennan*); 30 species of
ornamental plants (*Rhodendron*, growing in the higher mountains, *Paeonia lutea*, *Rosa odorata*, *Aesculus wangli*); 18
species medicinal plants (e.g. *Magnolia officinalis*, *Eucommia
ulmodes*, *Aconitum* spp. and *Scopolia carniolicoides*); wild
species of crops (e.g. *Oryza*. spp., *Malus sikkimensis* and
Mangifera sylvatica).

CONSERVATION OF THE PLANT RESOURCES

Two major methods are being used for plant resource conservation:

1.   Nature reserves - Rare and endangered species are
being conserved in numerous scattered nature reserves.   There
are 30 reserves in Yunnan; 3 are at the national level.   The
largest one is Xishuangbanna National Nature Reserve, established in 1986, with an area of two million hectares.   Species
of tropical seasonal rainforest vegetation are the object of
main protection (Xu, 1988).

2. Cultivation and conservation in botanic gardens - Since
1985, 50 species of rare and endangered plants have been successfully transplanted into botanic gardens, utilizing information gained from studies of the habitats and physiology of
these species.   The techniques of plant tissue culture has
been widely applied in the rapid propagation of these plants
[e.g successful propagation of *Alsophila spinulosa* (Chen *et
al.*, 1991).

EXPLOITATION OF GENE RESOURCES FROM WILD PLANTS

Considering the needs for the improvement in the genes of
crops, especially in the resistance to pathogens by genetic
engineering, a research project in Kunming Institute of Botany
was initiated in 1986 to screen for anti-fungal proteins and
their genes in wild plants.   An anti-fungal protein (GAFP) of
14kD has been isolated and characterized from the corm of *Gastrodia elata* (Hu *et al.*, 1988).   This protein was shown to be
the most important substance in the defense mechanism of the
corm, toxic specifically to fungi species.   From the seeds of
Phytolacca, an antifungal polypeptide (PAFP) has been isolated
and the amino acid sequenced (Hu *et al.*, 1991) protein has
been identified and discovered to have high activity in inhi-
biting the growth of pathogens *Fusarium oxysporum* f. *vasinfectum* and *Verticillium dahliae* in cotton and *Fusarium grami-
nearum* in wheat.   The isolation of the genes for these two
proteins is being done at the present time.

Literature Cited

Chen, Z. Y., F. L. Zhang, Q. Y. Lan, Z. F. Xu and G. D. Tao. 1991. Studies on the propagation and conservation of germplasm in *Alsophila spinulosa* (in Chinese, English summary). *Acta Botanica Yunnanica* 13: 181–188.

Hu, Z., X. Z. Liu, Q. Z. Huang, X. M. Zhu, and D. Yang. 1991. Isolation, characterization and amino acid sequencing of an antifungal protein from the seeds of *Phytolacca americana*. (in English). In: *Proceedings of the Chinese Symposium on Polypeptide* (1990, Shanghai). Science Press, Beijing. (in press)

Hu, Z., Z. M. Yang and J. Wang. 1988. Isolation and partial characterization of an antifungal protein from *Gastrodia elata* corm (in Chinese, English summary). *Acta Botanica Yunnanica* 10: 373–380.

Kunming Institute of Botany. 1984. *Index Florae Yunnanesis* Vol. 1-2. The Peoples' Publishing House, Yunnan.

Xu, Z. F. 1988. The conservation and utilization of tropical plant germplasms in the tropics of Yunnan (in Chinese, English summary). *Acta Botanica Yunnanica* 1: 113–124.

CURRENT STATUS OF PLANT GENETIC RESOURCES IN ETHIOPIA

Abebe Demissie

Plant Genetic Resource Centre/Ethiopia
P.O. Box 30726, Addis Ababa, Ethiopia

Summary - Ethiopia is considered as one of the richest genetic resource centers in the world. Primitive cultivars/landraces of several major crops *viz.* wheat, barley, sorghum, field pea, faba bean, coffee, etc. and wild relatives of some of the world's important crops are abundant in the Ethiopian region. This potentially useful resource is under constant threat of depletion. Cognizant of the importance of conserving genetic diversity and in order to avert the danger of genetic erosion, Plant Genetic Resources Centre/Ethiopia (PGRC/E) was established in 1976 to collect, conserve, evaluate, document and promote the utilization of crop plant germplasm occurring in Ethiopia. The PGRC/E, since its establishment has mounted a series of expeditions to collect and conserve the diversity in crop plants occurring in Ethiopia. As a result 19,248 accessions were collected by the Centre alone. In general, currently the PGRC/E holds 50,150 accessions of 99 species obtained through collection, repatriation and donation. A great portion of the material has been evaluated for various characteristics at appropriate agro-ecological sites. The material collected over the years is being conserved using various storage practices depending on the storage behavior, type and the nature of the species.

INTRODUCTION

Ethiopia is considered as one of the richest centers of genetic resources in the world. Wide altitude and temperature ranges, high humidity and extreme forms of rainfall patterns coupled with complex topography make Ethiopia a major region of genetic diversity for many crop plants. It is believed that indigenous crops such as teff, *Eragrostis tef* (Zucc.) Trotter; "noog", *Guizotia abyssinica* (Cass.); and ensat,

Ensete ventricosum (Welw.) Cheesm. were first domesticated in
Ethiopia. Numerous major crops species including wheat (*Triticum durum*), barley (*Hordeum vulgare*), sorghum (*Sorghum
bicolor*), sesame (*Sesamum indicum*), castor (*Ricinus communis*),
coffee (*Coffea abyssinca*), etc. are also known to have their
center of diversity in the Ethiopian region (Vavilov, 1951).

The recognition of the presence of an enormous genetic
wealth in the country led to the establishment of the Plant
Genetic Resources Centre/Ethiopia (PGRC/E) in 1976 in order to
collect and save these dwindling resources from extinction.
Since then, several major and minor collecting expeditions
covering a wide range of agro-ecological zones in the country
have been undertaken.

COLLECTION

The Plant Genetic Resources Centre/Ethiopia (PGRC/E)
holds about 50,150 accessions (Table 1) of mainly cereals and
millets (75.30%), pulses (9.50%), oil crops (7.92%), stimu-
lants (2.41%), spices (2.07%), root crops (0.52%), medicinal
plants (0.29%) and others (1.99%) (Table 1) (Worede, 1983).
The collection consists of 99 species of 99 genera which are
known to possess certain economic and social values. Well
over 38% of the current holding has been collected by PGRC/E
since its establishment in 1976. The remaining germplasm was
obtained through repatriation (7.21%), donation (4.26%) and
previous collection by national institutions (27.55%). A substantial portion of the collection (22.67%) represents samples
selected out of original population accessions with mixed
species and apparent morphological differences.

CONSERVATION

Ideally all plants should be conserved as evolving populations in their natural ecosystems. However this is not
practically feasible for all species. Biologists have, therefore designed a number of ways to overcome such difficulties
that can be broadly divided into two categories: *ex situ* and
in situ; the former is the major conservation strategy
employed by PGRC/E.

Table 1. Current Crop/plant Germplasm Collection at PGRC\E

Crop\Plant Category	Total Collection Accession	%	PGRC\E collection Accessions	%
Cereals/millet	37,761	75.30	11,002	57.17
Pulses	4,766	9.50	3,448	17.91
Oil Crops	3,973	7.92	2,440	12.68
Stimulant mainly(coffee)	1,208	2.41	344	1.79
Spices	1,034	2.06	805	4.18
Root Crops	262	0.52	262	1.36
Medicinal Plant	150	0.29	150	0.78
Others	996	1.99	797	4.12
Total	50,150	100.00	19,248	100.00

EX SITU CONSERVATION

In the Ethiopian genebank the major conservation method employed is the seed bank. The Center holds an active collection for research and distribution of seeds and a base collection for long-term security storage. Orthodox seeds are first dried to 3-7% moisture in a dehumidified drying room operating at 15-20% relative humidity and 18-20°C before they are packed in aluminium foil bags. Seeds meant for immediate use and multiplication are stored in the temporary storage at +4°C and

the long term material is kept in the base genebank operating
at -10°C.

Recalcitrant seeds and species that do not readily produce seeds are conserved *ex situ* in field genebanks. The coffee accessions, root crops such as yam (*Dioscorea bulbifera*),
'Oromo dinich' (*Coleus edulis*) and spices *viz.* ginger (*Zingiber officinale*), Korrorima (*Afromomum corrorima*), etc. are
conserved at appropriate agro-ecological areas in field genebanks. These are maintained in close proximity to the major
research establishment in order to allow the utilization of
these materials in national research programs. This operation
represents the conventional conservation strategy employed by
PGRC/E.

IN SITU CONSERVATION - LOOSELY TERMED

In Ethiopia, farmers play a pivotal role in the conservation of landraces as they hold the bulk of genetic resources.
Peasant farmers always retain some traditional seed stock for
security even at difficult times unless circumstances dictate
otherwise. A landrace conservation and enhancement program,
which provides a unique opportunity to conserve and develop
traditional seed materials that are adapted to the local agro-
ecological realities, has been embarked upon, in strategic
sites under conditions where extensive environmental stresses
prevail. It was primarily designed to help the Ethiopian
farmer maintain this crop diversity by protecting major cultigens from extinction while improving the genetic perfor-
mances of such materials. This communal conservation strategy
which is undertaken at the grass-root level involving farmers,
scientists and extension works has been in progress since 1988
(Worede, pers. comm.).

The Center, in implementing its conservation task, has
employed an integrated use of community and conventional methods of conservation as indicated above. However new/advanced
techniques such as tissue culture, cryopreservation, etc.,
have only been envisaged.

CHARACTERIZATION/EVALUATION OF GERMPLASM

Evaluation of genetic resources is of prime importance in

making a large collection available for subsequent utilization. Results of preliminary evaluation can assist user communities to select the right materials for breeding. The work
in this regard focuses basically on recording those agro-
morphological traits which are highly inheritable, can easily
be seen by the eye and expressed in all conditions with no
significant change in different environments. In this
case, evaluation means rating on mainly physiological or morphological traits that are easy to evaluate and also stable at
different sites and from year to year.

Early studies in the screening of crop germplasm were
undertaken by various scientists. These works indicated that
particular characteristics that distinguish some Ethiopian
germplasm from that of other origins have been observed.
Numerous breeders have also discovered some highly desirable
genetic characteristics in Ethiopian landraces. These
include disease resistance (Qualset, 1975), high lysine content in barley (Munck, 1972), and rust resistance in wheat
(Creech & Reitz, 1971) for a few accessions of Ethiopian germplasm collections. For instance, Qualset reported that 2% of
the world collection of barley, showed virus resistance and
that all resistant strains were of Ethiopian origin.

In recent years numerous research activities have been
carried out at PGRC/E to determine the potentialities of the
traditional cultivars. This included studies on low pH tolerance, drought resistance and constituent analysis. Krauss and
Ghiorgis (1986) studied 654 wheat accessions from the PGRC/E
collection for low pH tolerance. According to their finding
8.7% of the total germplasm tested were tolerant to acid
soils. Recently the protein content of 500 wheat germplasm
(Demissie et al., 1989) and 500 accessions of sorghum were
analyzed using a Near Infrared Reflectance (NIR) Analyzer.
These studies identified wheat and sorghum lines with protein
contents as high as 22% and 15% respectively on a dry matter
basis.

Currently screening for drought tolerance on various crop
types is under way using a rapid screening technique employing
root-shoot ratio. Encouraging results have been obtained.

AGRO-MORPHOLOGICAL EVALUATION

A considerable number of germplasm samples has been evaluated for various agro-morphological characteristics since the
establishment of the Centre. Hitherto 21,731 accessions of
cereals, 4,388 accessions of pulses, 5,819 accessions of oil
crops that account for 75% of the general holding have been
characterized for various traits (Table 2). This study
revealed that most of the wheat germplasm is early maturing
with relatively small vegetative and floral organs (Mekbib &
Habtemariam, 1989). High variation in chickpea for some agronomic traits and high yielding faba beans were also identified. Preliminary investigations have also revealed the
existence of early maturing, lodging and disease resistant,
and high grain quality sorghum strains in local collections.
Similar studies undertaken on castor indicated the presence of
early maturing and short-stemmed types in local collections.

PATHOLOGICAL EVALUATION

The traditional crops are attacked by numerous fungal and
bacterial diseases in Ethiopia. In recent years, the PGRC/E,
in collaboration with national and international institutions
has carried out germplasm evaluation for resistance to various
diseases. Some of the major studies include the joint
PGRC/EICRISAT - Addis Ababa University evaluation work on some
1,700 sorghum accessions for tolerance/resistance to bacterial
leaf streak and stalk borer, and screening of durum wheat
strains for resistance to stem, stripe and leaf rust in collaboration with the Scientific Pathobiological Institute, Ambo.
The latter study included 502 genotypes and 392 accessions
were found to be resistant to one, two or three of the major
rust diseases.

UTILIZATION

Some of the collected germplasm resources have been utilized very effectively in plant breeding in recent years in
this country. The considerable amount of human power, investment and funding that go into collection and preservation of
genetic resources can only be rationalized by the successful
utilization of the resources in plant breeding for the welfare

Table 2. Number of accessions characterized and agro-
morphological traits recorded for each crop type*

| Crop type | No. of accessions | | No of traits |
	Total	Characterized	considered
Arachis hypogaea	9	17	27
Brassica spp.	1,178	1,929	25
Carthamus tinctorius	185	173	30
Cicer arietinum	851	1,300	12
Coffea arabica	1,151	4,224	29
Eleusine Coracana	1,394	617	19
Eragrostis tef	3,328	703	18
Guizotia abyssinica	1,023	1,975	22
Heliantus annuus	67	24	21
Hordeum vulgare	12,954	6,783	18
Lablab purpureus	36	11	36
Lathyrus sativus	268	177	15
Lens culinaris	495	684	11
Linum usitatissimum	1,025	1,236	18
Pennisetum typhoides	143	22	25
Phaseolus spp.	313	157	36
Pisum sativum	1,330	1,067	12
Ricinus communis	505	68	23
Sesamum indicum	466	397	27
Sorghum bicolor	7,444	6,966	23
Trigonella foenum-			
graecum	487	320	13
Triticum spp.	11,823	6,440	17
Vicia faba	1,422	630	10
Vigna unguiculata	49	23	36
Total	47,266	35,943	

* Some of these accessions have been characterized more than
once.

of the whole of mankind.

Many local collections are already incorporated into the
national crop improvement programs through the national yield
trials and are even utilized in specific areas of breeding
including those related to resistance and adaptation (Worede,
1988). Promising genotypes of native noog (*Guizotia abyssinica*), Ethiopian mustard (*Brassica carinata*), linseed, and
safflower collections have been advanced to the different levels of replicated variety trials. Over the past years considerable entries of sorghum derived from Ethiopian landraces
have been recommended for release for various agro-ecological
zones in the country.

On an international scale, Ethiopian material, especially
wheat, has been known ever since the expedition by N.I. Vavilov in the 1920's. The significance of these materials to
breeders in USSR has been substantial. Much of the germplasm
was also made available to European wheat breeders. Konzak
(1977) reported five Ethiopian durum wheat entries with a high
content of protein, lysine and other amino acids. These lines
made up 42% of the highly utilized strains of durum in American wheat breeding programs. The source of high lysine and
high basic amino acids for durum wheat were exclusively from
Ethiopia (Negassa, 1986). The Ethiopian sorghum germplasm has
been found useful in cold tolerance, high protein (lysine),
good grain quality, disease resistance and diversity, as indicated by its use in the U.S. sorghum conversion program
(Kebede, 1986).

REGIONAL/INTERNATIONAL COOPERATION
   Training course on genetic resources activities -PGRC/E
and cooperating national institutions have hosted and actively
participated in a series of training courses on genetic
resources in the last two years. The first two sessions were
sponsored by Unitarian Service Committee of Canada (USC). The
beneficiaries included NGO staff working in the field of agriculture and development-related fields coming form various
African and Asian countries. Recently a regional training
workshop on evaluation and utilization of genetic resources of
local crops of agricultural importance in Africa was organized

at PGRC/E.   The training workshop was co-sponsored by FAO,
UNEP and CTA.

Germplasm Exchange and collaborative links - Ethiopia
adheres to the principle of free exchange of germplasm with
bona fide users in accordance with its national policy.   Germplasm is sent to or exchanged with foreign countries as long
as mutual advantage in such an exchange exists.   It is generally based on a three-point contact policy whereby Ethiopian
breeders, scientists from cooperating institutions and PGRC/E
are involved.   Exemplary collaborative work in this regards
include ICRISAT for sorghum, millet and chickpea; ICARDA for
pulses, wheat and barley; IRRI for rice.   A list of germplasm
donated and/or exchanged is shown in Table 3.

At the international level, collaborative ties exist with
IBPGR, FAO, ICRISAT, ICARDA, IRRI, IITA and similar international centers.   Likewise, PGRC/E maintains working links with
regional and national genebanks such as Gatersleben Genebank,
Forschungsantalt fuer Ladwirtschaft (FAL) in Brauschweig, Germany; Bari Germplasm Institute, Italy, The USSR Genetic
Resource System among others (Worede, 1986).

CONCLUSION

From this account it can be seen that the Ethiopian
crop/plant germplasm has specific attributes of utility in
breeding efforts.   The landraces have desirable traits that
can be crossed with high yielding improved lines.   The traditional cultivars, being adapted to diverse agro-ecological
conditions soil and water stress, can provide useful genes for
introducing tolerance to drought, low pH, cold and diseases.

Collections of root crops, medicinal plants, weedy
species and wild relatives of cultivated species are relatively scanty.   Existing collections for some species are so
small that an effort to collect and protect the wild species
of crops both for in situ and ex situ conservation should be
strengthened.   Until now, little or no significant work has
been undertaken in forest resource conservation and this is
one area where an obvious gap exists despite an apparent environmental degradation in the country.

Appropriate conservation techniques, especially for

Table 3. List of germplasm dispatched to various institutions
         and/or countries*

Species	No. of accessions	Organization/ Country	Purpose
Capsicum spp.	68	CATIE	Exchange
Oryza sativa	3	Philippines	Exchange
Triticum spp.	50	Sweden	Research
Brassica spp.	15	"	"
Hordeum vulgare	5	"	"
Eragrosits tef	5	"	"
Guizotia abyssinica	6	–	Exchange
Sesamum indicum	4	Tanzania	"
Hordeum vulgare	20	–	Research
Cicer arietinum	104	ICRISAT	Collector's sample
Cajanus cajan	14	"	" "
Sorghum bicolor	2	"	" "
Vicia faba	9	IBPGR	Research
Pisum sativum	10	"	"
Triticum spp.	18	"	"
Hordeum vulgare	54	"	"
Brassica spp.	29	"	"
Linum usitatissimum	15	"	"
Hordeum vulgare	55	"	"
Guizotia abyssinica	2	Sweden	"
Vicia faba	1	"	"
Lupinus spp.	5	"	"
Lens culinaris	79	–	"
Solanum incanum	4	IBPGR	"
Phaseolus spp.	6	Italy	"
Lablab sp.	6	"	"
Guizotia abyssinica	82	U.K.	"
Eleusine coracana	35	"	"
Ocimum sp.	1	Switzerland	"
Canavlia sp.	4	Italy	"
Eragrostis tef	62	Israel	Exchange

Table 3 (continued)

Species	No. of accessions	Organization/ Country	Purpose
Eleusine coracana	50	Burundi	"
E. coracana	70	IBPGR	"
Sorghum bicolor	20	U.K.	"
Oryza spp.	16	Philippines	Research
Eragrostis tef		Germany	"
Sorghum bicolor		"	"
Sorghum bicolor	24	Canada	"
Vernonia galamensis	6	Zimbabwe	Exchange
Brassica spp.	15	China	Research
Hordeum vulgare	69	U.K.	"
Triticum spp.	5	"	"

Total   1048

* This list does not include the collector's sample (107 accessions mainly cereals and pulses) that was dispatched to China and 216 accessions of cereals and millets to Europe. The latter was for research work by PGRC/E staff members in U.K. and Denmark.

---

plants which do not produce seeds, are inadequate and need improvement. These kinds of conservation techniques will have to be initiated in view of the economic and practical advantages of conservation. Priority should be given to the conservation of ecological and economic key plant species threatened by extinction. Genetic, species and ecosystem diversity conservation should be encouraged through an integrated approach that unites local farmers, scientists and governmental agencies, as well as non-governmental agencies.

Literature Cited

Creech, J. C. and L.P. Reitz. 1971. Plant germplasm now and for tomorrow. *Adv. Agron.* 23: 1-49.

Demissie, A., G. Habtemariam, and R. Feyissa. 1989. Ethiopian Wheat Germplasm Collection, Conservation Characterization/Evaluation and Utilization. Pp. 327-334. In: D. G. Tanner, M. Van Ginkel, W. Mwangi (eds.), *The sixth regional wheat workshop for Eastern, Central and South Africa.* Addis Ababa, Ethiopia, October 2-6, 1989.

Kebede, Y. 1986. The role of Ethiopian sorghum germplasm in the national breeding program. Pp. 223-231. In: J. M. M. Engels (ed.), *The conservation and utilization of Ethiopian Germplasm.* Proceeding of International Symposium Addis Ababa, Ethiopia. 13-16 October 1986.

Konzak, L. 1977. Genetic control of the content, amino acid composition, and processing properties of proteins in wheat. *Advan. Genet.* 19: 407-528.

Krauss, A. and H. M. Ghiorgis. 1986. Further studies on tolerance Ethiopian wheat germplasm to acid soil. *PGRC/E-ILCA Germplasm Newsletter* 8: 7-11.

Mekbib, H. and G. Habtemariam. 1989. Ethiopian Wheat Germplasm Evaluation and Utilization. Paper presented at the International workshop of Genetic Resources in Wheat Improvement, May 17-23, 1989. Aleppo (ICARDA) Syria.

Munck, L. 1972. High lysine barley - a summary of the present research development in Sweden. *Barley Genetic Newsletter* 2: 54-59.

Negassa, M. 1986. Pattern of diversity of Ethiopian Wheat (*Triticum* spp.) and a gene centre for quality breeding. *Plant Breeding* 97: 147-162.

Qualset, C. D. 1975. Sampling germplasm in a Centre of diversity: an example of disease resistance in Ethiopian barley. Pp. 81-94. In: O. H. Frankel and J. G. Hawkes (eds.), *Crop genetic resources for today and tomorrow.* Cambridge University Press, Cambridge.

Vavilov, N. I. 1951. The origin, variation, immunity and breeding of cultivated plants. *Chronica Botanica* 13: 1-366.

Worede, M. 1983. Crop genetic resources in Ethiopia. Pp.
143-147. In: J. C. Holmes and W. M. Tahir (eds.), *More
food from better technology*. FAO, Rome.

_____. 1986. An Ethiopian Perspective on Conservation and Uti-
lization of Plant Genetic Resource. Pp. 197-211. In: J.
M. M. Engels (ed.), *The conservation and utilization of
Ethiopian germplasm*. Proceeding of International Symposium, Addis Ababa, Ethiopia, 13-16 October 1986.

CONSERVATION OF PLANT DIVERSITY IN JAPAN

Kunio Iwatsuki

Botanical Gardens, University of Tokyo
Tokyo, Japan

Summary - An overview of the status of floristic work and efforts to conserve the flora of Japan is presented. Factors involved in plant extinction are discussed.

INTRODUCTION

The Japanese Archipelago is situated at the eastern end of Asia. The topography of this archipelago is complex and the Black Current washes the eastern coast of it. Influenced by warm and humid climate, Japan has a very rich flora; more than 5,000 species of vascular plants are found in its small area covering only 378,000 square kilometers.

FUNDAMENTAL INFORMATION ON PLANT DIVERSITY

The flora of Japan has been studied fairly well since the first *Flora of Japan* was prepared by C. Thunberg in 1784. The most recent treatment in English was published in 1965, translated from Jisaburo Ohwi's *Flora of Japan* originally published in 1953, in Japanese. This flora discusses all the vascular plant groups, but unfortunately the Ryukyus and the Bonins are not included. Two books have been published recently on the Ryukyu Islands: a book in Japanese by Hatusima (1971) and one in English by Egbert Walker (1976). These are good complements to Ohwi's *Flora of Japan*, published both in Japanese and English. The modern treatment of the Bonin flora is not available; the most recent compilation is by Toyoda (1981), only in Japanese.

Besides these important contributions, there are a variety of books published in Japanese: various colored illustrations and many local floras. Revisional works on a variety of plant groups have been published and our information on the diversity of Japanese flora has been greatly expanded in recent years. *Illustrations of Wild Flowers of Japan* (Hara et

al., 1981-89) was published in five volumes during 1981-89,
and a volume of the ferns and the allied plants is now in
press. These volumes include detailed description of each
taxon at the rank of species, genus and family, keys to these
taxa and includes all the taxa native to Japan including the
Bonins and the Ryukyus. Colored pictures taken in the field
are given for most of the species. Based mainly on these volumes and including all the information available, a concise
flora of Japan is under preparation and will be published in
five volumes very soon. All the results obtained through the
biological research on Japanese species are to be included in
the flora.

The recent development in biotechnology has made the study
on plant diversity possible to include such analysis as cyto-
taxonomy, comparative physiology, adaptive biology, molecular
systematics,, etc. Biosystematic analysis on Japanese plants
has been made intensively, and it is impossible to enumerate
such works in a restricted space. Important results from such
biological studies are incorporated in compiling the flora and
our knowledge of plant taxa is growing day by day.

BIOSYSTEMATIC STUDIES ON PLANT DIVERSITY IN JAPAN

As the fundamental observations of flora of Japan have
been made comparatively well, systematic studies of a variety
of taxa are being made by applying various biological techniques. Developing the classical karyological research on
wheat by Kihara (1924) cytogenetical analysis of Japanese
plants has been actively conducted. Chromosome numbers have
been counted for most of the Japanese plants, and the information on karyotypes, polyploidy, hybridization, morphology of
each chromosome, etc. is available.

The description of a variety of characteristic features,
including the data on chemical substances, has been made, and
our knowledge of the Japanese plants is increasing steadily.
However, systematic botanists expect to have complete data on
each taxon. Thus, there are a lot of matters to be analyzed
and observed.

Reproductive adaptation is one of the features now under
observation in relation to adaptive strategy of the vegetative

world. Fundamental observations of reproductive types of each taxon will be made, and theoretical research will be combined with actual available data.

All the biosystematic studies are now under promotion, especially in the recent field of molecular systematics. Cytogenetic analysis is modernized with the information available on isozyme analysis as well as on DNA information. Observations have been made very recently on both chloroplast and nuclear DNA on various taxa of Japanese plants. The complete sequences of chloroplast DNA have been published for tobacco (Shinozaki et al., 1986), rice (Hiratsuka, oral communication) and *Marchantia polymorpha* (Ohyama et al., 1986). Additional complete DNA chloroplast sequences for various species are being made and the data is expected to be reported soon.

Systematic interests are beginning the comparative study of DNA information as the basic data is becoming available. *Adiantum capillus-veneris* chloroplast DNA (cpDNA) is in a clone bank and is being used as heterologous probes for systematic studies of the leptosporangiate ferns (Hasebe and Iwatsuki, 1990a, b). Nuclear DNA of the fern *Asplenium cataractarum* was studied by Murakami et al. (1990) and compared with that of the flowering plants. Thus, molecular systematic studies are being conducted in Japan for comparing DNAs.

RARE AND ENDANGERED SPECIES IN JAPAN

Although the Japanese Archipelago is covered with the green revolution, this was developed by recent human activities. Nature is greatly influenced by modern agriculture, and the flora is changing, especially in the areas where the population density is higher. To record the changing aspects of the flora of Japan, a research group was organized to make a survey of rare and endangered species in Japan. This group was sponsored by WWF-Japan and Nature Conservation Society of Japan, and supported by Japan Society of Plant Taxonomists. A preliminary list of rare and endangered vascular plants of Japan was published in 1989.

The list enumerates 899 species under the IUCN categories of Extinct, Endangered, Vulnerable and Unknown. Among 5500

species of vascular plants native to Japan, 16.4% are threatened species, which is approximately one-sixth of all the
flora of Japan (Table 1).

Table 1. Rare and Threatened Species in Japan

Species	Extinct	Endangered	Vulnerable	Unknown
Pteridophytes	10	11	74	6
Gymnosperms	0	0	4	0
Monocotyledons	10	54	271	23
Choripetalae	10	41	190	3
Sympetalae	5	41	138	4
Total	35	147	677	36

Based on field observations, there appears to be three
reasons why so many species are threatened:

1. development of nature by human activity,

2. over-collection of wild species, mostly for trade for
horticultural purposes, and

3. general change of the natural environment influenced by
obvious or unknown reason by local and/or global factors.

It is of great concern to note that the control of the
above factors is hardily possible at the moment. However, we
should note that many Japanese people are becoming aware of
these facts.

The dynamics of the changing flora are under observation
by botanists as well as amateur plant observers, and the
changing aspects are recorded by the Agency of Environmental
Protection. The dynamics of each threatened species is
observed and analyzed in various ways. Many endangered
species are brought into the botanical gardens and ex situ
conservation of threatened species is one of the functions of
several Japanese botanical gardens. A data-base of living
collections in botanical gardens is now in preparation.

One of the activities along this line is expected to be
the recovery of native sites of endangered species. The Bonin
Islands are typical oceanic islands and have been developed in
the past hundred years. Thus, many endemic species are

seriously endangered. Some threatened species (most of them
hardly cultivated) have been propagated in the botanical gardens and planted in the native cites with great collaboration
of the people. It is expected that the native flora there
will be kept in its original condition, through it is necessary to observe it carefully for many years (Shimozono & Iwatsuki, 1986).

Literature Cited

Hara, H. *et al.* (ed). 1981-89. *Wild Flowers of Japan. I,
    Herbaceous Plants - Monocotyledoneae* (1982). *II, Herba-
    ceous Plants - Choripetalae* (1982). *III, Herbaceous Plants
    -Sympetalae* (1981). *Woody Plants I -Gymonspermae and Cho-
    ripetalae* (1989). *Woody Plants II - Sympetalae and Monoco-
    tyledoneae* (1989). Heibonsha, Tokyo.
Hasebe, M and K. Iwatsuki. 1990a. Chloroplast DNA from *Adian
    tum capillus-veneris* L., a fern species (Adiantaceae);
    clone bank, physical map and unusual gene localization in
    comparison with angiosperm chloroplast DNA. *Curr. Genet.*
    17: 359-364.
_____ and ____. 1990b. *Adiantum capillus-veneris* chloroplast
    DNA Clone Bank: as useful heterologous probes in the
    systematics of the Leptosporangiate ferns. *Amer. Fern.
    Jour.* 80: 20-25.
Hatusima, S. 1971. *Flora of the Ryukyus.* Okinawa Biol.
    Assoc., Naha (In Japanese).
Kihara, H. 1924. Cytologische und genetische studien bei
    wichtigen Getreiden Arten mit besonderer Ruechsicht auf
    das Verhalten der Chromosomen und die Sterilitaet der Bas-
    tarden. *Mem. Coll. Sci. Kyoto Imp. Univ.* B., V., 1:
    1-200.
Murakami, N., Y. Tanaka, K. Takishima, Y. Minobe, M. Matsuoka,
    S. Kiyota, S. Hatanaka and K. Sakano. 1990. Structure of
    cDNA clones for the small subunit of Ribulose-1, 5-bis-
    phosphate carbosylase/oxygenase from a fern (*Asplenium
    cataractarum* Rosenstock, Aspleniaceae), and its comparison
    with those of various seed plants. *Bot. Mag. Tokyo* 103:
    419-434.

Ohyama, K., H. Fukuzawa, T. Kohchi, H. Shirai, T. Sano, S.
    Sano, K. Umesono, Y. Shiki, M. Takeuchi, Z. Chang, D.
    Aotoa, H. Inokuchi and H. Ozeki. 1986. Chloroplast gene
    organization deduced from complete sequence of liverwort
    Marchantia polymorpha. *Nature* 322: 572-574.
Ohwi, J. 1953. *Flora of Japan*. Shibun-do, Tokyo (In Japanese).
_____. 1965. *Flora of Japan*. Smithsonian Inst., Washington,
    D. C.
Shimozono, F. & K. Iwatsuki. 1986. Botanical gardens and the
    conservation of an endangered species in the Bonin
    Islands. *AMBIO*. 15: 19-21.
Shinozaki, K., M. Ohme, M. Tanaka, T. Wakasugi, N. Hayashida,
    T. Matsubayashi, N. Zaita, J. Chunwongse, J. Obotata, K.
    Yamaguchi-Shinozaki, C. Ohto, K. Torazawa, B. Y. Meng, M.
    Sugita, H. Deno, T. Kamogashira, K. Yamada, J. Kusuda, F.
    Takaiwa, A. Kato, N. Tohdoh, H. Shimada and M. Suguira.
    1986. The complete nucleotide sequence of the tobacco
    chloroplast genome: its gene organization and expression.
    EMBO Jour. 5: 2043-2049.
Thunberg, C. P. 1784. *Flora Iaponica sistens plantas insula-*
    rum Iaponicarum.
Toyoda, T. (ed.). 1981. *Flora of Bonin Island*. Aboc Co., Ltd.,
    Kamakura (In Japanese).
Walker, E. 1976. *Flora of Okinawa and the Southern Ryukyu*
    Islands. Smithsonian Inst., Washington, D. C.

STATUS OF FOREST CONSERVATION FOR MAINTENANCE
OF BIODIVERSITY IN NIGERIA

Zac. O. Gbile

Forestry Research Institute of Nigeria
Private Mail Bag 5054, Ibadan, Nigeria

Summary - The different uses of plants are discussed.
Every plant is created for a purpose. These is need for constant rejuvenation of all plant species including food
crops. *In situ* and *ex situ* conservation are the only means of
maintaining gene reservoirs. The vegetation zones in Nigeria
tend to be delineated by adaphic conditions particularly
drainage and biotic factors. Nigerian forests contain over
560 species of trees which attain a height of at least 12 m,
ferns and orchids are present in wet areas. The dominant
species of each vegetation zone is discussed. Nigeria has
only 10-11% of land area under reserved forest estates. There
are at present 12 established Strict Natural Reserves. In
cases where it is not possible to maintain an adequate genetic
base for any particular endangered ecosystem or species in
areas where it occurs, *ex situ* conservation is appropriate
(pollen storage, seed storage and clone banks). Limitations
of *in situ* conservation in the arid zone is discussed and the
roles played by the Forestry Research Institute in combating
desertification are enumerated. Different measures for conservation have been taken by the Federal Government and State
Forestry.

INTRODUCTION

The conservation of biological resources has received considerable attention in the past decade and is now the subject
of unprecedented debate by national and international technical and political bodies (Lyman, 1984). Plants supply us with
food, drugs, building and other raw materials, fuel, fibers
and ornamentals. Some years ago, only very few species were
thought to be economic, notably the mahoganies and some white
wood such as obeche. Recently, the number of economic species

has increased three-fold. No plant can really be considered
useless even if it has not been the subject of a thorough
study, since it may have value in the future in as yet a quite
unknown field. Tropical moist forests contain many world relatives of modern food crops: rice, millet, yam, banana, etc.
Species of intensive domestication require constant 'rejuvenation' with fresh germplasm in order to resist new types of
diseases and pest, environmental stresses and also to increase
productivity and nutrient contents. Genetic improvement in
plant species is usually based on the remaining natural
genetic variation (gene pool) within the species. Lost
species and those with very poor genetic diversity cannot be
reconstituted if germplasm has not been conserved earlier.
Conservation areas (*in situ*) and those conserved *ex situ* are
the only means of maintaining the gene reservoir of our natural heritage of plants in all of their primitive diversity to
meet man's present and future needs (Kio *et al.*, 1985).

Nigeria is a very large country with estimated land area
of 90,879,138 hectares and extensive vegetation cover. Vegetation resources is just one of the country's natural renewable resources that require conservation (sustained management
approaches) so that the country can derive maximum benefit
from them. The other renewable resources include wild animal
resources, water resources, marine and fisheries resources and
soil resources. With the exception of marine and fisheries
resource, the forestry section (forest services) in Nigeria
are involved in the conservation of all the resources mentioned above.

FOREST AND FORAGE RESOURCES OF NIGERIAN NATURAL FORESTS

In a general way, the vegetation zones in Nigeria corre-
spond to climatic zones, but tend to be delineated by edaphic
conditions, particularly drainage and biotic factors. The
distribution of vegetation depends chiefly on the fact that
the climate tends to become drier with increasing distance
from the sea. The operative factors affecting the distribution of species are the seasonal rainfall and the duration and
degree of atmospheric desiccation during the dry season.

The major vegetation formations are the mangrove forest

and coastal swamps, freshwater swamps, lowland rain-forest
(high forest), derived savanna (merging into Southern Guinea
Savanna), Northern Guinea Savanna, Sudan Savanna, Sahel, Bauchi Plateau, submontane forest and grassland (Keay, 1959).
Four-fifths of the land area bear savanna with the remainder
being mainly a mosaic of farms and forest.

The Nigerian forests contain over 560 species of trees
which attain a height of at least 12 m (ca 40 ft) and a girth
of 60 m (ca 2 ft). A wide range of life-forms, including many
epiphytic ferns and orchids are present, particularly in the
wetter forests, where grasses are rare.

Sandy beaches, coastal creeks and lagoons stretch along
the coast, flanking a 25,900 sq. km. (10,000 sq. miles) area
of mangroves and swamps. The typical soil of the mangrove
swamps is a blue alluvium that is discolored to brownish-grey
at the surface. It supports a close, single-storyed stand of
trees with stilt-roots and often pneumatophores. The mangrove
swamp is dominated by *Rhizophora* spp. (*R. racemosa*, pioneer
species; *R. mangle* and *R. harrisonii*) and *Avicennia africana*.
The coastal vegetation consists of forest and thicket with
Chyrosbalanus icaco and *Thespesia populnea* tending to be
dominant species.

In the freshwater swamps (e.g. in Mamu Forest Reserve),
the main canopy is rather open and in the gaps, dense tangles
of shrubs and lianes form an almost unpenetrable undergrowth
(Keay, 1959). Climbing palms with hooked spines are characteristic, while *Anthostema aubryanum*, *Symphonia globulifera*,
Allanblackia floribunda and *Raphia* spp. are prominent.

Inland from the coastline, the tropical rainforest (lowland "high forest") prevails to a depth of from 95 km (60
miles) to 160 km (100 miles) northwards. The lowland forest
covers an area of 103,600 sq. km. (ca 40,000 sq. miles).
There is no single canopy layer; crowns exist at all levels,
with discontinuous taller emergents, to give an almost completely irregular structure. The families Leguminosae (e.g.
Brachystegia, *Gossweilerodendron*), Meliaceae (*Entandrophragma*,
Guarea, *Khaya*, *Lovoa*), Moraceae (*Antiaris*, *Melicia*),
Sapotaceae (*Aningeria*, *Chrysophyllum*, *Manilkara*, *Mimusops*),
Sterculiaceae (*Cola*, *Mansonia*, *Pterygota*, *Sterculia*, *Triplo-*

chiton) and Ulmaceae (*Celtis, Holoptelea*) are particularly
well represented in this zone.

The derived savanna (merging into the Southern Guinea
Savanna, see Fig. 1) zone stretches across Nigeria from east
to west separating the rainforest to the south from the fire-
swept Guinea savanna in the north. Derived savanna vegetation
has been induced by the activities of farmers in clearing
land, thereby admitting a grassy ground-layer which encourages
fire. The mean annual rainfall in this zone is about 1,440 -
1,780 mm (ca 56-70 inches) and the dry season lasts for about
three months. Oil palms (*Elaies guineensis*) are a characteristic relict in derived savanna. The common woody species

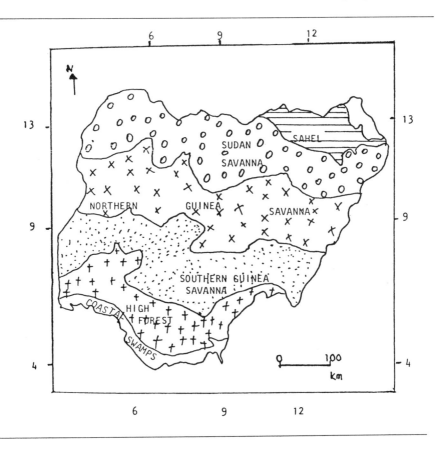

Figure 1. Vegetation map of Nigeria (after Keay, 1959)

are *Daniellia oliveri*, *Dialium guineense*, *Albizia adianthifo-
lia*, *Lophira lanceolata* and *Hymenocardia acida*. Characteristic trees of the relict forest are *Celtis zenkeri*, *Cola gigan-
tea*, *Anthonotha macrophylla* and *Treculia africana*. The common
grasses in this zone are species of *Andropogon* and *Loudetia
arundianaceae*.

To the north of the valleys of the Niger and Benue Rivers
savanna predominates. Undisturbed savanna is characterized
mainly by low and open woodland in which the tree crowns generally do not form a close canopy, and by a ground-cover of
coarse grasses which is virtually complete. The grasses usually have rolled-up leaves and are perennials; they are burned
annually and often form tussocks. At the northern limit of
the derived savanna, the vegetation takes on the character of
Guinea savanna with fringing forest in the river valleys. The
mean annual rainfall in the Southern Guinea Savanna zones is
between 1,140 mm and 1,520 mm (ca 45-60 inches), and the dry
season lasts for about four to five months. The common woody
species in this zone are *Daniellia oliveri*, *Diospyros
elliotii*, *Brysocarpus coccineus*, *Drypetes floribunda*, *Ceiba
pentandra*, *Entada abyssinica*, *Brachystegia eurycoma*, *Nauclea
latifolia*, *Fagara zanthoxylum*, *Burkea africana* and *Prosopis
africana*. The most abundant grasses in this zone are species
of *Andropogon*, *Ctenium* and *Hyparrhenia*.

North of the limits of the Southern Guinea Zone is the
Northern Guinea zone (Fig. 1). The mean annual rainfall here
is between 1,020 mm and 1,140 mm (ca 40-45 inches), and the
dry season lasts for five to six months. The common woody
species in this zone are *Isoberlinia doka*, *Albizia zygia*,
Anthoclesta vogelii, *Cola hispida*, *Erythrophleum suaveolens*,
Annona senegalensis, *Manilkara obovata* and *Ficus exasperata*.
The common grasses in this zone are species of *Schizachyrium*,
Loudetia, *Ctenium* and *Andropogon*.

The most northerly regions of the country, extending
southwards for approximately 250 km (155 miles) from the
northern border are loess-derived desert soil of the Sudan
savanna and also containing extensive areas of seasonal swamps
with black cracking soil (fadama). The mean annual rainfall
in the Sudan zone is between 510 and 1,140 mm (ca 20-45

inches) and the dry season lasts for five to seven months.
The common woody species in this zone are *Acacia nilotica* var.
nilotica, *Maerua crassifolia*, *Acacia senegal*, *Afrormosia
laxiflora*, *Afzelia africana*, *Isoberlinia doka*, *Vitellaria
paradoxa* and *Burkea africana*. The most abundant grasses in
this zone are species of *Aristidia*, *Brachiaria*, *Panicum*,
Chloris and *Digitaria*.

At the northeastern limit of the Sudan Savanna and border-
ing Lake Chad in the Sahel zone, the mean annual rainfall is
approximately 250 mm (ca 10-20 inches) and the dry season
lasts for seven to eight months. The common woody species in
this zone are *Acacia laeta*, *A. tortilis* subsp. *raddiana*,
Commiphora pedunculata, *Cordia sinenesis*, *Grewia tenax*,
Leptadenia pyrotechnica and *Salvadora persica*.

Within the northern Guinea zone there are a few small isolated areas of tropical rainforest-lowland rainforest developed on an accumulation of peat. The soil consists of a hori-
zon of grayish brown clay peat, overlying an exceedingly plastic gray clay with sparse orange mottles. In this type of
forest, the canopy is of uniform height (about 15 m or 50
feet) and without emergents. It is dominated by *Ficus umbel-
lata*, *Syzygium guineense*, *Voacanga thouarsii* and *Cleistopholis
patens*. Between the canopy and the ground is a diffuse irregular understory of *Ficus leprieurii*, *Psychotria psycho-
trioides*, *Bridelia micrantha* and *Xylopia aethiopica*.

Nigeria's highlands consist of the Jos Plateau, Obudu Plateau, Mambilla Plateau, Vogel Peak Massif and Gotel Mountains.
All rise to over 1000 m and carry peculiar floras. They are
not regarded as belonging to any of the main vegetation zones.
The annual rainfall generally is higher than it is in the surrounding lowland and this is probably due to orographic
effects according to Hall (1971).

PRESENT STATUS OF CONSERVATION

The present status of conservation was discussed by Kio
and Gbile (1990). When the Nigerian Forestry Department was
set up in 1897 (Adeyoju, 1975), its duty included protection
of the existing natural high forest and savanna ecosystems
(formation) and drawing up plans for reforestation of denuded

lands. Both the forest and savanna formations are sources of
numerous products that sustain both the rural and industrial
economy of the country. The official government policy was to
constitute 25% of the natural forest as forest reserves which
should be conserved and managed on sustained yield basis so as
to derive maximum benefit at all times from the plant, animal,
water, soil and other resources obtainable from the forests.

At present the country has been able to bring only 10-11%
of the land area under reserved forest estates. Appendix A
shows the amount of land in reserved forests under the various
current land use practices (Allen, 1981). It is unlikely that
the area of reserved forests will exceed this amount in the
near future because of the ever increasing pressures from
other land use sectors of the country's economy. Already substantial areas of reserved forest land have been lost to other
competing land uses which include urban development, agriculture, road and oil pipeline construction. Some of the forest
estates are also being converted to game reserves and parks
for wild animal conservation.

IN SITU CONSERVATION

In situ conservation is feasible where pressure on the
natural ecosystem is light and techniques of perpetrating the
ecosystem or species are unknown. Strict Natural Reserves,
Game Reserves, National Parks, Botanical Gardens and Arboreta
or manager forest reserves will serve as adequate means of
conserving such ecosystems or species.

There are at present 12 established Strict Natural
Reserves in Nigeria (Fig. 2)...and the list of the protected
plots have been given (Ola-Adams and Lyamabo, 1977; Ola-Adams,
1977). In cases where it is not possible to maintain an adequate genetic base for any particular endangered ecosystem or
species in areas in which it occurs, ex situ conservation is
appropriate. Seed storage, pollen storages, establishment of
clone banks and the use of tissue and meristem culture in
vitro techniques will serve as adequate means of conserving
such species.

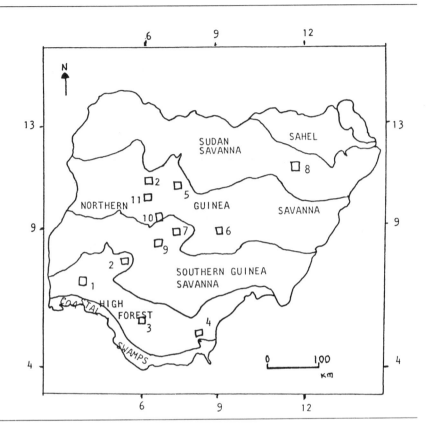

Figure 2. Distribution of Natural Reserves (established conservation areas) in Nigeria (SNR nos. 1-12).

NATIONAL PARKS AND GAME RESERVES

In Nigeria, the establishment of the Yankari Game Reserve
and the Kainji Lake National Park was the start of serious
efforts at the conservation of the endangered species of animals and plants. For example, a variety of vegetation types
is found in the Kainji Lake National Park forming a mosaic of
wooded savanna interspersed with patches of woodland and riverine vegetation. Within this park is a wide variety of animals, such as elephants, hippopotamus, wart-hog, buffalo,
roan, hart beast, bush buck kob, red buck, water buck, red-

flanked ducker, grey ducker, oribi, lion and leopards. The primates are represented by Tanalus monkey, red patas monkey and baboon. Bird life is also abundant.

BOTANIC GARDEN AND NATIONAL HERBARIUM

Many plant species in the Nigerian vegetation are fast diminishing and perhaps on the brink of extinction. It is, however, generally acknowledged that the most effective method of saving such species is to maintain and preserve the complete habitat in which they grow. Rare species could be cultivated in a safe area (botanic gardens) as well, which has the same or nearly the same habitat or environmental conditions as their natural abodes.

There is an arboretum at the Forestry Research Institute of Nigeria with quite a diverse and varied tree species, some originating outside Nigeria. A few of our universities (i.e. University of Ibadan and Obafemi Awolowo University) have botanic gardens for purely educational purposes.

There is at present no national botanic garden or herbaria. It is only the Forest Herbarium which is national in scope. The establishment of a national botanic garden and national herbarium at Abuja are in future plans.

The Botanic Garden of the University of Ibadan contains over 25,000 plants (herbs, shrubs and trees) which are mainly indigenous with a few exotic species. Since this garden is situated in the dry, high forest zone and coupled with its teaching aim, the spread of plants is limited, but a few plants are in conservation. The Forest Herbarium at the Forestry Research Institute of Nigeria has about 130,000 collections of preserved plants, most of which are indigenous. The international recognition gained by this herbarium is evidenced by the numerous requests for its services; cooperative national and international investigations are undertaken from time to time.

EX SITU CONSERVATION

The maintenance and regeneration of species in their original or natural occurrence (in situ) has been described. The conservation of genetic resources outside their origin (ex

situ) includes pollen and seed storage, clone banks, clone
tissue or meristem culture (*in situ*) for storage under artificial conditions.

POLLEN STORAGE

Systematic investigations on pollen storage for conservation started in the last century, Zirkle (1935) reported that
transportation of pollen of the date palm took place as early
as 2400-2000 B.C. In Nigeria, pollen collection for storage
to conserve the genetic variability of native forest species
started only recently. This procedure arose as a result of
the realization of the need for maintaining the gene pool of
the natural forest and to enable controlled crosses in tree
breeding as it is often difficult to make crosses at the time
of natural pollen maturation and dehiscence. Pollen materials
of *Tectona grandis* and *Pinus caribea* (exotic) have been collected and stored on individual and provenance groupings for
breeding and physiological investigation. Efforts on native
species under the West African Hardwoods Improvement Project
(WAHIP) have concentrated on high forest species, such as *Triplochiton scleroxylon*, *Terminalia ivorensis*, *Terminalia superba* and *Mansonia altisssima*. Pollen storage for conservation
started for *Triplochiton scleroxylon* in 1975 as the need to
store viable pollen of this species was recognized as a vital
aspect of its improvement and genetic conservation particularly because of its selective and intensive exploitation for
wood (export) and for domestic use which was drastically reducing its genetic base in Nigeria.

With constant, annual pollen available in some amount,
detailed pollen studies have been possible (Leakey, *et al.*,
1981). At present pollen from 85 sources have been stored.
These mainly come from the southwestern high forest zone of
Nigeria. Appreciable viability has also been achieved for
these collections with storage at -17°C and the maintenance of
low humidity (3-5%) within the storage tubes.

SEED STORAGE

Germplasm conservation by seed storage is a very valuable
aspect of gene conservation. However, in Nigeria, the lack of

adequate seed storage conditions and the rapid loss of seed
viability (Olatoye, 1968) are common problems in genetic conservation of tropical species by seed storage. At present,
efforts at the Forestry Research Institute now allow for adequate storage of seeds annually. The central seed store, a
major seed collection and distribution center in Africa, now
stocks seeds of indigenous species and seeds of exotic forest
species. Although efforts of the central seed store is geared
towards plantation establishment, specific efforts at the conservation of important forest species by seed storage are
being made, particularly pollen storage of *Triplochiton scler-*
oxylon under WAHIP. For storage, seeds are usually dried in
the air-conditioned laboratory environment (28 + 2°C) where no
drastic temperature fluctuations can affect seed viability.
In most cases, a dehumidifier is employed to reduce humidity
and hasten seed drying. At between 6 and 8 % seed moisture
content, seeds are placed in air-tight kilner jars with some
silica gel to keep moisture content at the desired level. For
short term storage, the seeds are stored in the cold room at
0.5°C, whereas, for the long term storage, they are placed at
-17°C in a freezer. The latter has proved most successful
with recent evaluations indicating 73% viability in seed
stored in 1976 despite inclement power supply. At this low
temperature, it is indeed possible to conserve seeds for
appreciable periods.

Long term seed storage of other species has also started
in the West African Hardwood Improvement Project (WAHIP).
These include *Terminalia ivorensis, Mansonia altissima, Termi-*
nalia superba, Ceiba pentandra, Chlorophora excelsa and
Daniellia ogea.

At present, seeds from 580 sources have been stored for *T.*
scleroxylon, while an average of 10 have been stored for the
other species. *T. scleroxylon,* being self-sterile, seeds need
to be collected and stored continuous if adequate germplasm
conservation is to be made. This is to salvage the remaining
genetic materials available in this species, as much as wild
pollination allows.

CLONE BANKS

Insufficient seed for plantation development led to the
use of a viable method of vegetative propagation. In the case
of *Triplochiton scleroxylon*, this resulted in the shift in
emphasis from provenance trials to the more specific and
genetically, productive method of clonal trials, based on
vegetatively produced plants. Propagules have been planted
out since 1975. These large scale field trials (0.3 million
trees sited at 5 experimental stations) are laid down in a
series of replicated experiments, initially concerned with
interpreting differences between and within clones, single-
parent seed lots and interactions within site variations. The
experiments have been designated to become clonal gene banks
after evaluations are completed. Apart from the above, progenies and straight clonal gene banks are established annually
at Onigambari Forest Reserve and Ore Station near Ondo in
southwestern Nigeria.

Genetic resources of *Triplochiton scleroxylon* and *Terminalia ivorensis* have been conserved by other methods of vegetative propagation. Budding and successful grafting of materials from mature but selected trees (provisional plus trees)
have been executed at the Forestry Research Institute of Nigeria. 330 budded and grafted stocks have been maintained for
over 10 years in the nursery. Apart from the potted plants
serving as conservation stands, they also provide materials
for flower induction and the consequent improvement of this
species when breeding facilities and opportunities come. They
were the basis of the present knowledge on the breeding systems of this species.

LIMITATIONS OF *IN SITU* CONSERVATION IN THE ARID ZONE

In Nigeria, the arid zones spread over six extreme northern states (Sokoto, Kaduna, Katsina, Kano, Bauchi and Borno)
which lie north of latitude 12°N. These are areas which were
once rich in vegetation. These arid zones are characterized
by long dry seasons, which could last for about seven months
or more with an effective rainy season of approximately 100
days. These areas are also characterized by wide range of
seasonal variation in temperature and relative humidity. In

addition to this, the areas support relatively high population
density - man and livestock animals (Ogigirigi, 1985).
The areas on the zones previously referred to as Forest
Reserves are being threatened with desertification. As a
result, most of the natural vegetation is now disappearing.
The few surviving trees are often deflated (Fig. 3). The
effect is more serious in Sokoto and Kano States. The
increasing demand for wood and farmland results in over-

Figure 3. *Combretum glutinosum* (Deflated). The only surviving
indigenous tree around Gidan Kaura (Sokoto State, Nigeria).

exploitation of forest resources without commensurate efforts
to replenish. Sand dunes, in addition to destroying vegetation, has also rendered many people homeless, especially in
Sokoto State (Fig. 4). About 75% of the territorial land area
of Kano State has also been engulfed by desert-like conditions. According to FAO/UNEP (1981), the inhabitants of Kano,
a city in the Nigerian savanna, crops 75,000 tons of firewood

every year from the surrounding woodlands. This is another
example where desert is being formed at a quicker rate than
anywhere in the world.

Figure 4.   Sand dune at Gidan Kaura (Sokoto State).

The most effective way in curbing desertification and
restoring plant and animal life is through reforestation,
either by establishing shelterbelts, increasing tree densities
as scattered-farm trees or mixed or monocultural stands, home
or farm boundary tree planting or by block planting or plantation. Among these alternatives, the establishment of shelterbelt is the most effective since this involves planting of
rows of trees across the prevailing wind direction; hence the
shelterbelts prevent wind damage at the leeward side and
improve the microclimate conditions. The Shelterbelt Research
Station, Kano, was set up in 1975 by the Forestry Research
Institute of Nigeria to investigate the problem. The role of
the research station is purely to conduct research and develop
suitable methods and techniques for establishment and management of appropriate vegetation cover in the semi-arid areas of
Nigeria as well as to determine their effects on the environment. The project areas span the entire semi-arid zones of
Nigeria - all areas north and latitude 12°N extending from

Sokoto State through Kaduna, Kano, Bauchi to Borno State.

The Forestry Research Station in its measure to conserve the vegetation and agricultural crops of the affected areas has recorded the following achievements:

(a) Field trials of over 23 species in different locations throughout the semi-arid zone indicate that *Azdirachta indica* (Neem), *Eucalyptus camaldulensis*, *Cassia siamea* and *Dalbergia sissoo* are well adapted for survival and good growth on various locations.

(b) Social and economic surveys conducted in some villages situated within or near shelterbelt indicate that the benefits occurring from the present of shelterbelt include:

1) Stability of villages as against continued migration from villages without shelterbelt schemes.

2) Enhanced economic position of villages due to increased agricultural productivity from their farmlands and improved vegetation cover.

3) Very pronounced amelioration of micro-climate in the villages due to abundance of shade provided by shelterbelt trees and natural vegetation.

4) Complete disappearance of the frequent sand storms which were causative agents to the disappearance of many plant species.

CONCLUSION

The Federal Government of Nigeria recently inaugurated a National Advisory Committee to look into the problem of conservation in the country and to recommend from time to time measures to be taken to achieve the required goal. *In situ* and *ex situ* conservation are encouraged even at the local government level. Traditional healers who are involved in total removal of medicinal plants are encouraged to establish their own medicinal plant gardens. Annual burning by farmers is often discouraged and the slogan "Plant a Tree" is the order of the day in Nigeria.

In an attempt to combat future consequences associated with desertification, the Kano State Government has recently taken the following steps:

(a) Establishment of tree nurseries in almost all the district councils of the states.

(b) Development of "protective" forest shelterbelts.

(c) Establishment of industrial, fuelwood plantations and water catchment planting along the state water sources.

With above measures taken in the desert encroached zones, conservation of natural vegetation will be achieved.

Literature cited

Adeyoju, S. K. 1975. *Forestry and Nigerian economy*. Ibadan University Press, Ibadan, Nigeria.

Allen, P. E. T. 1981. Land use in Nigeria. Paper presented at Forestry Management Planning Seminar, Ibadan, March, 1981.

FAO/UNEP. 1981. Tropical forest resources assessment project. Forest resources of tropical Africa. Part II. Country Briefs, FAO,Rome, Italy.

Hall, J. B. 1971. Environment and vegetation on Nigeria's Highlands. *Vegetatio* 23: 339-359.

Keay, R. W. J. 1959. *An outline of Nigerian vegetation*. 2nd Ed. Federal Government Printer, Lagos, Nigeria.

Kio, P. R. O. and Z. O. Gbile. 1990. Practical aspects of forest conservation for maintenance of biodiversity of sub-Sahara Africa-the Nigerian Experience. Invited paper for the International Conference on Conservation of Tropical Biodiversity "In Harmony with Nature", The Malayan Nature Society and the Ministry of Science, Technology and Environment, Kuala Lumpur, Malaysia, June, 1990.

_____, Z. O. Gbile, B. A. Ola-Adams, E. O. Bamgbala and D. O. Ladipo. 1985. Strategies for the conservation of natural forests and grazing lands in Nigeria.

Leakey, R. R. B., N. R. Fergussin and K. A. Longman. 1981. Precocious flowering and reproduction biology of *Triplochiton scleroxylon* K. Schum. *Comm. For. Rev.* 60(2): 177-126.

Lyman, M. T. C. 1984. Plant genetic resources. *Newsletter* 60: 3-4.

Ogigirigi, M. A. 1985. An approach to development of forest resources of the Sudan-Sahelian zone of Nigeria. In: J. S.

Okojie and O. O. Okoro (eds.), *Proceedings of the 15th Annual Conference of the Forestry Association of Nigeria,* Yola, Nigeria.

Ola-Adams, B. A. 1977. Strategies for conservation and utilization of forest genetic resources in Nigeria. *Nig. Jour. For.* 1191: 32-40.

_____ and D. E. Lyamabo. 1977. Conservation of natural vegetation in Nigeria. *Environ. Conserv.* 4(3): 217-226.

Olatoye, S. T. 1968. Seed storage problems in Nigeria. Paper presented at the 9th Commission Forestry Conference, India.

Zirkle, C. 1935. *The beginning of plant hybridization.* University of Penn. Press, Philadelphia. 213 pp.

Appendix A SOURCE: P. E. T. Allen (1981)

LAND USE AND VEGETATION OF NIGERIA AS DETERMINED BY SLAR

Formation	Total Area (hectares)	Area in Forest Reserves (hectares)	% of Total in Forest Reserves	% Areas Reserved to Total Land Area
Forest	8,874,255	1,825,821	18.247	2.00
Woodland	4,197,209	702,041	7.016	0.77
Wooded Shrub Glassland/ Woodland Transition	23,747,396	4,106,254	41.037	4.50
Shrubland and Thicket	2,288,311	667.664	6.673	0.73
Grassland/ Shrubland Transition	1.779.382	88.600	0.885	0.10
Grassland	12,821,302	1,613,230	16.122	1.77
Farmland	35,870,552	874.285	8.737	0.96
Plantations and Agric. Projects	276,500	122.044	1.220	0.13
(Forestry Plantations)	(133,761)	(109,699)	(1.096)	(0.12)
Water, Rivers, Built-up Areas	1,024,231	6,252	0.063	0.007
	90,878,138	10,006,191	100.00	10.967

INDIGENOUS FLORA CONSERVATION IN KENYA

Titus K. Mukiama

Department of Botany, College of Biological and Physical
Sciences, University of Nairobi, P. O. Box 30197
Nairobi, Kenya

Summary - Kenyan vegetation has been as varied as its
natural habitats, although considerable environment degradation has taken its toll in recent years. The major loss of
the indigenous vegetation began with the introduction of plantation farming by European settlers early this century. Virgin forests on appropriated land were cleared away, the local
people were concentrated into tribal reserves, and a conflict
that spanned decades was thus started. The reserves were soon
overcrowded, overstocked and in excess of their carrying
capacity. A net illegal movement of the African population
into restricted areas ensued, and is continuing even today.
Forests were a favorite target area as they provided farmland,
building materials and fuel wood. In recent years, forest
hardwood species have become threatened as the demand for timber continues to rise. Rangeland species, too, have become
threatened as new settlements continue to infiltrate these
marginal areas. The Kenya government has, however, responded
to these problems sufficiently by taking both legislative and
administrative measures aimed at achieving a sustained status
of natural resources conservation. Consequently, environmental awareness and participation is gradually permeating into,
not only institutional and community, but also family levels
of society. While fully employing *in situ* means, *ex situ*
methods are likely to play a major role in the future, particularly as the country utilizes the new technologies emerging
from tissue culture, cell culture and genetic engineering.

INTRODUCTION

The fine layer of rock, water and air that surrounds
planet earth, along with the living organisms for which it
provides support is called the biosphere (Dasman, 1976). It

is the habitat of all the life forms on this planet. A biosphere change whether natural or induced by man, can bring
about different results. If a specific habitat is destroyed,
the species agglomeration within that habitat is lost. Alternatively, species from the lost habitat can regroup, albeit in
different combinations, in different habitats. A third possibility is that if no suitable local habitat exists then
species may become extinct. In general, this scenario
urgently requires remedial measures to reverse, or at least
repair the damage already incurred by the biosphere. Such
measures are often broadly referred to as conservation.

According to Jones (1987), conservation incorporates preservation, protectionism and conservation *per se*. Preservation
refers to the prevention of destruction, and is contemporarily
applied to extremely rare and valuable objects or biota. Protectionism is the defense of an object which has been overused
or misused in the past. In the extreme, it is synonymous with
preservation in that no change from an existing state is
allowed apart from maybe renovation to repair damage. Protection policy on a species or habitat or resource is usually
made to prolong its existence. For a non-renewable resource,
protection policy may be dictated mainly by economic factors.
Renewable resources however cannot be managed in such a single
deterministic manner. For biotic resources such as forests
and animal populations, the usable yield of the resource will
be a function of ecological factors such as age, sex ratio,
biotic potential, migration and fluctuations in the environmental circumstances like climate, soil fertility, pollution,
etc. Conservation embraces all of preservation and some of
the points relating to protection. It also must involve a
conceptual appreciation of the need to use biospheric
resources in ways which place the least restriction upon the
future well-being of organic and inorganic resources. In its
broadest sense, it is now involved with husbanding the
resources of the whole biosphere. Pettigrew (1982) adds that
a complete conservation policy involves the incorporation of a
conservation ethic into the everyday lifestyles in human
society. Instead of showing a myopic interest in economic
growth rates, we should be focusing upon the ability of

society to be self-supporting in terms of energy and material
use. Sandbach (1980) notes that for conservation to make
progress, its case must be presented in a form which is
directly understandable by non-scientific and anti-
conservationist bodies. It should not become an elitist attitude which can be afforded only by the most wealthy and best
educated. A successful conservation policy must address the
requirements of the biologist for absolute conservation and
also those of the industrialist, agriculturalist, peasant,
pastoralist and indeed the whole of humanity as they strive to
satisfy their own needs and those of the market. Taking into
account this broad conceptual framework of conservation this
article attempts to review the problems, progress and current
status of efforts to conserve the indigenous flora of Kenya.

HISTORICAL BACKGROUND

The history of conservation in Kenya, whether of the flora
or the fauna, is closely related to the coming of European
settlers into the country early this century, and their relationship with the indigenous African people. During pre-
colonial times, the local human populations lived more or less
in harmony with their environment since the degree to which
they exploited it rarely surpassed their daily needs. Significant losses of fauna and flora only occurred after natural
disasters like floods, droughts, fires, volcanic eruptions,
etc. Previously therefore, conservational concerns may have
been irrelevant since the numbers of organisms were under more
or less predictable natural regulation.

The main European settler groups arrived in Kenya at the
end of both the First and the Second World Wars. Much of the
higher land was appropriated on a wholesale basis from its
African owners and given or sold to settler planters. The
local people were confined to overcrowded reserves with infe-
rior soil qualities, and in some cases, these reserves were
taken from them later. Most of the appropriated land was virgin forest, which was cleared for the development of cash crop
plantations, mainly tea and coffee. Large acreages of adjoin-
ing woodlands and highland savanna were further declared for-
feit and allocated for dairy and beef production. Soon, the

'reserve' areas were overcrowded, over-stocked and fast deteriorating.

Patches of natural forest, covering only 3% of the total
land area were gazetted and protected under the Forest and
Timber Acts. These laws specified what and where to harvest
forest products, but paid no attention to conservation and
replacement of resources. Natural regeneration was presumed
to be a sufficient remedy. The rangelands, which cover over
three-quarters of Kenya's total land area, were partly
reserved for nomadic pastoral peoples (mostly in the northern
and southern districts), national parks and game reserves, or
appropriated for large-scale ranching.

In the post-colonial era, the same attitudes prevailed.
The agriculture policies pursued did not sufficiently address
agriculture as part of the totality of the ecology of the
entire region. They encouraged the increased acreage of cash
crops and exotic vegetation at the expense of indigenous crops
and natural vegetation that was of no immediate benefit. As
the human population explosion began to take its toll, official reaction was often too late, and usually to crisis situations. Large areas of rangelands gradually turned into semi-
desert, empty spaces began to appear in the midst of rain forests, riverine forests in several areas disappeared and some
rivers dried up. Weather patterns began to change.

THE NATURAL VEGETATION

Most, if not all, of the major natural vegetation types
and habitats on earth are found in Kenya, ranging from the
glaciers atop Mt. Kenya to the dry deserts of the north and
the tropical waters of the Indian Ocean. The major vegetation
types are described in detail by Lucas (1968), and only a
brief summary will be given.

The Afro-alpine glaciers and moorlands occur in areas over
3200m above sea level, and are found in the high mountain
areas of Mt. Kenya, Mt. Elgon, the Nyandarua and Mau ranges.
In areas above 2000m, the highland moist forests occur and are
characterized by *Podocarpus*, *Octea*, *Nuxia* and bamboo species.
The tropical rain forests which exhibit the richest diversity
of plant life are typified by the Kakamega forest in Western

Kenya. Genera such as *Celtis*, *Chrysophyllum* and *Croton* are
common. These tropical forests have been largely decimated,
although traces of similar lowland moist forest may be seen in
the coastal region interspersed with cultivation, with *Brachy-
laena* and *Manilkara* being the most common genera. The higher
mountain areas also do contain dry highland forests, which
are also found in many isolated hills throughout the country.
Typical genera in these forests include *Juniperus*, *Olea*,
Teclea and *Euclea* among others. A dry coastal forest, locally
known as the 'Miombo' woodland borders the thorn bushlands of
the interior. This forest is typified by the presence of
species of *Sterculia*, *Chlorophora*, *Memecylon* and the occa-
sional baobab *Adansonia digitata*. Also to be found in the
coastal vicinities is the palm woodland, typified by *Hyphaene*
and *Borassus* palms on open grassland and ground water sites,
especially in the south and north. Ground water and riverine
forests, which are supported by underground seepage of water
are common, especially in the northern, eastern, southern and
rift valley districts of Kenya. They include the Dodori, Boni
and Kitobo forests, while riverine forests occur along many
river valleys including the Tana, Ewaso, Nyiro, Kerio and
Turkwell. Mangrove forests flourish in several areas along
the Indian Ocean shores or the interface between brackish
waters, reefs and creeks. Species of *Avicennia*, *Ceriops*,
Rhizophora, *Sonneratia*, *Bruguiera*, *Heritiera*, *Lumnitzera*, and
Xylocarpus have been described.

Most of Kenya is covered by an extensive woodland-
bushland-grassland vegetation mosaic. True woodlands border
the highlands of central and western Kenya, but gradually
diminish to give way to bushland thickets and on to arid bush-
lands that border semi-deserts. Thorn woodlands are probably
derived from the forests and found in the immediate vicinity
of the forests from Ukambani to the Lake Victoria lowlands and
as far north as Samburu. The *Acacia* sp. predominate, princi-
pally *A. abyssinica*, *A. lahai* (in higher altitudes),
A. polycantha, *A. xanthophloea* and *A. tortlis* (in the lower
altitudes). Other woodland areas include most coastal inland
areas stretching from the Shimba hills in the south, north-
eastwards to Malindi. The main tree species found include

Brachystegia spiciformis, Afzelia quanzensis and *Julbernardia magnistipulata.*

Thicket and bushland areas consist of about 50% shrubs and small trees growing closely together. Trees are usually of bushy habit, branching from near the base with an average height of 3-9m. The ground cover is usually grass, but is severely limited by soil erosion in many locations. This type of vegetation covers most of northern, eastern and southern Kenya. In the north, it gradually grades down into semi-deserts and deserts around Lake Turkana. Most national parks and game reserves like the Amboseli, Tsavo, Samburu, Meru, etc. are located in these areas. Such arid thorn bushes are dominated by the *Acacia/Commiphora* complex. Semi-arid bush-lands in neighboring areas are typified by *Combretum* sp., low *Acacia* sp., *Terminalia* sp. and *Hyparrhenia* sp. A comprehensive directory of Kenya's trees and shrubs is available for reference on these and other species (Dale and Greenway, 1961; Gillet and McDonald, 1970). Extensive grasslands occur in Kenya, especially in the rift valley. They are found particularly on the Laikipia-Lerogi plateau, the Kapiti and Loita plains. As already indicated, grass also forms the principal undergrowth of the rangelands. The principal savanna or plains grasses belong to the genera *Themeda, Cynodon* and *Digitaria.* In the wooded grasslands of the high plateaus evergreen and semi-evergreen trees are found, particularly species of *Scutia, Olea, Acacia* and *Acokanthera.* Rocky desert stretches occur in northern Kenya in the areas east of L. Turkana, the best known being the Chalbi and Kaisut. They grade into large semi-desert terrains supporting dwarf shrubs and without adequate ground cover. The terrain is rocky or sandy. These features are common in Turkana, Isiolo and Marsarbit districts as well as most of the North-Eastern Province. Other habitats supporting unique vegetation include the permanent swamps such as the Yala and Lorian, river deltas such as the Tana, and seasonal inland lakes like Amboseli. In these habitats, species of *Papyrus, Echinochloa, Phragmites* and *Miscanthus* are common. Similar environments fringe the freshwater lakes of Victoria, Naivasha, Baringo and Jipe. Saline and alkaline lakes occur in the rift valley. These are Lake Tur-

kana, L. Bogoria, L. Nakuru, L. Elmenteita and L. Magadi.
They all support unique flora and fauna. Along the Indian
Ocean coast, stretches of bare sand and dunes occur, with
Diani and Malindi providing panoramic sandy beaches. The
intertidal zones, fringing coral reefs and offshore islands
support thriving aquatic communities of both flora and fauna.
Many species of phytoplankton, marine algae and marine angios-
perms abound.

LOSS OF FLORA

The loss of flora in Kenya has been most evident in forests and rangelands. Forest makes up to 2.9% of the total
land area, which is about 1.5 million hectares; (of this 52.5%
is natural forest, 23% consists of woodlands, plantation forests and bamboo account for 10.6% each, and the remaining 3.4%
are mangrove forests). Rangelands make up 77% of the total
land area. The extensive reduction of Kenyan forests that has
taken place over the last two decades is a direct result of
the high population increase which has tripled over that
period. This enormous population pressure has led to an
increase in subsistence agriculture that has not spared even
closed government forests. Hand in hand with this struggle to
increase food production has been the unchecked over exploitation of forest resources for timber and fuelwood. Charcoal
production has been particularly harmful in the rangelands,
where large areas have assumed semi-desert conditions. In
addition, large numbers of people have continued to move into
and settle in these marginal lands to cultivate and rise livestock. Overstocking has resulted in the lowering of the carrying capacity of the land, leading to accelerated deterioration. In some forests, particularly those in the central and
western highlands, large scale tea and coffee plantations are
slowly being expanded at the expense of virgin forest land.
Indigenous vegetation has been destroyed in similar areas to
make way for commercial forest plantations of introduced
species. This has resulted in extensive but undocumented loss
of genetic resources. In other areas, a misdirected attitude
of encouraging the planting of introduced fruit and tree
species has not helped the cause of conserving indigenous

vegetation. In the wetlands, especially those in the rift
valley and western Kenya, natural vegetation is being
destroyed for the creation of irrigated farmlands. The net
result of these activities is that out of the estimated 2000
species of trees and shrubs in Kenya, an unknown number may be
already lost, 5% are considered endangered, and 8% are now
rare. Of the herbaceous species, 20% may be endangered.

The loss of flora attributed to human activity has been
exacerbated by the low status of awareness of environmental
and scientific education. In other cases, low standards of
living and political rhetoric from leaders have led to general
apathy in the affected populations. Other factors contributing to the loss of flora in Kenya include natural disasters
occasioned by climatic changes, especially long droughts in
the northern districts. In the savanna, large scale seasonal
fires may contribute significantly to the suppression of certain plant species. In the national parks and game reserves,
wildlife, particularly elephants have contributed considerably
to the destruction of certain plant species especially in the
Tsavo National Park. This may not result in species loss *per
se*, but does result in general deterioration of the environment which may eventually lead to loss of flora.

CONSERVATION

A. *Policy* - In order to provide mechanisms for environmental conservation and management in Kenya, the government has
formulated an environment policy and has established an institutional framework for its adequate implementation. However,
this policy is not consolidated into one environmental policy
document *per se*, but rather, is expounded in separate ministe-
rial statements, development plans and in several cases in
statutory legislations. Official concern for environmental
problems was first expressed in the 1970's, when the government proposed both administrative and legislative steps to
solve them at the Stockholm conference on the environment. It
is during this period that the headquarters of the United
Nations Environment Programme (UNEP) was set up in Nairobi. In
the subsequent government development plan of 1984-88 concrete
measures were taken in such areas as environment monitoring

and assessment, desertification, pollution control, human settlements and environment education. Environmental law in
Kenya is administrative in nature, and is mainly concerned
with the regulation of the use of water, land, minerals, forests and the protection of wildlife. Most of the environmental statutes inherited from the colonial administration cen-
tered on the extent of exploitation of the existing natural
resources as the main objective. The protection and conservation of the environment was therefore not a priority issue.

From the foregoing, it is clear that the law specifically
drawn up for the conservation of flora does not exist. Rather,
there is a conglomerate of regulations both in common and
statutory law whose interpretation hopefully serves the purpose of the conservation of the flora. Most of this authority
is vested in the Agriculture Act, the Forestry Act and the
Land Control Act. Other relevant legislation includes the
Timber Act, the Seed and Plant Varieties Act, the Grass Fires
Act and the Water Act. More specific, however are the laws
establishing national parks, which among other things set
aside such areas for the propagation, protection and preservation of wild animal life and natural vegetation. The destruction or collection of flora in such areas is prohibited.
National (game) reserves are similarly established on the same
lines, but human utilization of the resources therein such as
fuelwood collection and livestock grazing are sometimes
allowed. One notable exception to this available collection
of environmental laws is the absence of legislation on environmental impact assessment. Such legislation is important
because it would provide for the assessment of damage already
incurred through bad projects, and also provide for expert
advice on possible adverse environmental effects prior to the
initiation of new projects. A most desirable goal however
would be the formation of a complete law of conservation based
on sound ecological considerations.

B. *Target flora* - The forests form the main target plant
community for conservation efforts, mainly because of the
over-exploitation of important indigenous tree species for
timber. These include Meru oak (*Vitex keniensis*), Mukumari
(*Cordia africana*), Mvule (*Chlorophora excelsa*), Mahogany

(*Afzelia cuanzensis*), the East African olive (*Olea hochstet-
teri*), cedar (*Juniperus procera*) and podo (*Podocarpus milan-
jianus*).

In the rangelands, many slow-growing *Acacia* trees have
been destroyed for the supply of charcoal and firewood. The
sparse undergrowth of grasses and herbs has also been
destroyed through overgrazing. Since it would take a very
long time to efficiently re-plant the indigenous acacias, the
afforestation efforts under way are substituting those with
fast growing, drought-resistant leguminous tree species which
have to be specially developed through controlled selection
programs. Researchers and environmentalists are also getting
concerned with the possible loss of important but rare plant
species that are not necessarily found concentrated in any one
particular habitat. These plants include the indigenous fruit
trees and vegetables (Kokwaro, 1990), and species of chemical,
medicinal, pharmaceutical and traditional interest (Mukiama,
1989).

C. *Strategies and activities* – In Kenya, the overall
approach to conservation is environmental, with the flora
being considered both as a provision for ground cover to prevent desertification and also as an ecological component necessary for climatic stability. Secondary to this concept is
the concern for species threatened by overexploitation, especially the forest hardwoods and a multitude of other trees,
herbs and shrubs utilized as energy sources. These ideas certainly serve the cause of environmental conservation, although
they run short of addressing the issue of genetic resource
conservation, which is concerned with the conservation of as
many plant characteristics as possible from the myriads of
genomes around us. A start has been made though, and the
objective categories of the flora for conservation efforts
have sufficiently been identified as indigenous trees, fruits
and vegetables.

The main strategy for the conservation of forest species
is *in situ*, whereby designated indigenous forests are closed
to public exploitation by legal notice. The important timber
species mentioned earlier are protected in this way in the
major forests of Mt. Kenya, Nyandaruas, Mau, Kakamega, Mt.

Elgon and others. Other species of concern are conserved by *ex situ* methods, whereby collections of seed, or vegetative explant are made from areas of origin and planted in a central area where regeneration and protection can be effected.

In situ conservation activities are usually carried out by the Forest department, which is in charge of all forests in the country. Alternatively, individual farmers may exercise their own discretion and preserve indigenous vegetation within the boundaries of their land. *Ex situ* conservation activities are much more widespread, and form part of community and institutional activities. Community involvement is rather notable in the cultivation of wild fruits and some vegetables, particularly because the local people share a traditional knowledge base and are the likely beneficiaries of such programmes. They also have a social-cultural attachment to some of these species, and are therefore more likely to appreciate the objectives of the conservationists efforts. Most partici- pants in these community-based conservation programs are rural people who have grown up in the affected environments and can readily recall the circumstances prior to the destruction of the natural vegetation. Indigenous food crops are important retainers of primary gene pools, and are already providing materials for breeding programs.

The government of Kenya has directly involved all adminis- trative districts in environmental conservation activities through its District Focus Circular No. 2/88 of April 1988. Through it, environmental conservation committees were for- mally established in the districts, including their organiza- tional set-up and terms of reference. The main contribution of these committees is the annual tree planting exercises (which take place in May) which have resulted in the planting and regular maintenance of millions of new trees.

Non-governmental organizations have also been important contributors to the afforestation program. Two Kenyan bodies, the Greenbelt Movement and the Kenya Energy and Environment Organizations (KENGO) have made outstanding contributions to conservation in general and to indigenous tree planting in particular. The Greenbelt Movement is responsible for the afforestation of large tracts of land in several parts of the

country, while KENGO concentrates on the development of
indigenous tree nurseries and the documentation of indigenous
vegetable, and fruit tree species. KENGO has also been
involved in the education of the public on environmental
issues through seminars and the publication of posters, pam-
phlets, booklets and magazines. In 1988, KENGO organized a
very successful national expedition in which environmental
experts and media representatives undertook a study tour of
western Kenya and made recommendations to the government on
urgent issues concerning germplasm and environmental conservation (KENGO, 1989). International organizations like the
United Nations Environment Programme (UNEP) and the International Centre for Research in Agroforestry (ICRAF) have also
participated mainly in the training and education of environmental officers and members of the public at various levels.
Public institutions have also contributed significantly in *ex
situ* conservation activities. The Kenya Agricultural Research
Institute (KARI) recently set up a gene bank, whose functions
include germplasm exploration, collection, multiplication,
rejuvenation, characterization, evaluation and storage. The
Kenya Forest Research Institute (KEFRI) conducts research in
forest tree species and maintains a forest seed center. The
Department of Resource Surveys and Remote Sensing (DRSRS) col-
lects data on livestock and vegetative biomass in Kenyan
rangelands, assesses forest cover and maintains an ecological
data bank. Two herbaria, the University of Nairobi and the
National Museums retain specimens of up to 5 million plant
species from Eastern Africa, and plans are under way to establish a botanical garden. At the University of Nairobi, the
Department of Botany, Chemistry, Crop Science and Pharmacy,
among others, are involved in research on various indigenous
plant species, including clonal propagation through tissue
cultures. Similar projects are supported by the National
Council for Science and Technology (NCST) at other research
institutions and national universities.

CONCLUSION

Government action in the 1970's and 1980's supplemented by
general world reaction against the destruction of natural

vegetation, has set a foundation for continued conservation activities in Kenya. Education of the public and the allocation of sufficient resources will have to be maintained in order to achieve a sustained level of environmental protection and regeneration. Activities such as rural afforestation, nursery development, agroforestry and similar community-based programs must continue. In order to protect the limited natural resources, population growth must be controlled to manageable levels. Better planning for available local resources and the encouragement of international cooperation must continue. In addition, the available legislation regulating the exploitation of natural plant resources must be strictly enforced, although the formulation of a comprehensive law to that effect will continue to be the preferred objective.

In the long term, the country must continue to search for and evaluate emerging technologies that may present more efficient ways of plant conservation. These include the continued use of seed banks and the introduction of cell and tissue culture technology. Plant genomes could also be stored under appropriate conditions as plant tissue, DNA library clones or as molecular DNA itself.

Literature Cited

Dale, I. R. and P. J. Greenway. 1961. *Kenya trees and shrubs*. Government Printer, Nairobi, Kenya.

Dasman, R. F. 1976. *Environmental conservation*. Wiley, New York.

Gillet, J. B. and P. G. McDonald. 1970. *A numbered checklist of trees, shrubs and noteworthy lianes indigenous to Kenya*. Government Printer, Nairobi, Kenya.

Jones, G. E. 1987. *The conservation of ecosystems and species*. Croom Helm, London.

KENGO. 1989. *National expedition on genetic resources and habitats*. KENGO, Nairobi, Kenya.

Kokwaro, J. O. 1990. Seasonal traditional fruits and vegetables of Kenya. *Mitt. Inst. Allg. Bot. Hamburg* Band 23b s. 911-928.

Lucas, G. L. 1968. Kenya. Pp. 152-166. In: I. and O. Hedberg (eds.), *Conservation of vegetables in Africa south of the Sahara*. Almquist and Wiksells Boktryckeri AB, Uppasala.

Mukiama, T. K. 1989. Wild grains and neglected crops. *Resources J*. 1: 22-23.

Pettigrew, W. 1982. *Conservation*. Hodder and Stoughton, Sevenoaks, Kent.

Sandbach, F. 1980. *Environment ideology and policy*. Blackwell, Oxford.

TASK GROUP REPORTS

After the presentation of the status of species conservation efforts in various countries and new reports on *in vitro*
biotechnology for preservation, the participants of the DNA
Bank-Net meeting were divided into Task Groups to work on consensus statements for various concerns of the organization.
Following a half day meeting, a preliminary report was drafted
by the rapporteur, entered into a computer, and copies made
available for consideration by the entire congress. After an
open discussion, the rapporteur for each group amended the
draft report to reflect additional consensus. Occasionally,
no consensus could be reached, even among this rather small
congress of delegates. When this occurred, it is noted in the
group report.

GROUP A: RESOURCE IDENTIFICATION
Group Members: Bart Panis - Rapporteur, Daniel Abbiw,
L. A. Barreto de Castro, Zheng Sijun and Phillip Stanwood

**Task 1. Define the functions of working (dispensing) and
reserve (base) nodes in the DNA Bank network.**
The group recommended the following functions:
Working (dispensing) nodes:
a. Collection of plant material by taxonomists. This may
be the primary function of a particular node or be in association with other organizations such as universities, botanic
gardens, etc.
b. DNA extraction by molecular biologists or trained
staff.
c. DNA analysis/ gene replication by molecular biologists
or trained staff.
d. Distribution of DNA (genes, gene segments, oligonucleotides, etc.)
Reserve (base) nodes:
a. Long term DNA preservation in liquid nitrogen and
monitoring of potential DNA degradation.
b. Act as genetic reserve buffer for working nodes.
c. Replenishment of DNA if a working node experiences the

catastrophic loss of storage parameters and DNA.

Figure 1 depicts the relationship between working and
reserve nodes. Note the projected flow of plant materials and
DNA through the working (dispensing) node. It is likely that
some of the working nodes would be actively acquiring and/or
dispensing DNA from some geographic area (ex. Africa), yet
maintain separate cryovats, functioning as a reserve (base)
node for another area (ex. South America).

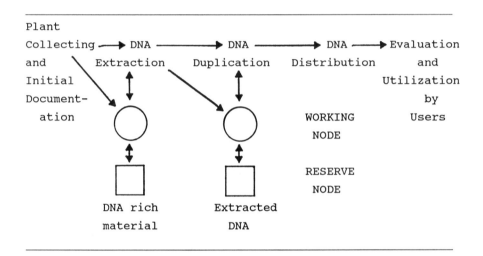

Figure 1. Schematic representation of the flow of materials
and the relationship between working (dispensing) and reserve
(base) nodes.

Considerable discussion was held on the problems of combining a working node function with a reserve node function.
The consensus appeared to be that if a reserve node is located
in conjunction with a working node that there be provision for
separate administration and in at least different rooms or
different buildings.

**Task 2: What are the general requirements for nodes in the DNA
Bank Network?**

The task group felt that the following were the minimum

requirements for nodes:
Working (dispensing) nodes:
 Personnel:
  Taxonomists/collectors
  Biochemists/molecular biologists
  Technicians for practical work
  Capable administration
 Equipment:
  Storage facilities (liquid nitrogen, cryovats)
  Extraction facilities (centrifuges, gel electrophoresis,
   UV spectrophotometer, etc.)
  DNA Analyses and PCR duplication (PCR thermal cycler,
   micro-centrifuges, etc.)
  Distribution system (packaging and mailing supplies)
  Computer (data base for inventory and correspondence)
Reserve (base) nodes:
 Personnel:
  Technicians
  Capable Administration
 Equipment:
  Storage facilities (liquid nitrogen, cryovats)
  Computer (data base for inventory and correspondence)
The group and congress felt that it was important that a work-
ing (dispensing) node have scientists who keep up with latest
developments, because molecular biology is rapidly changing.

Task 3: How many back-up (reserve) collections should there be for each DNA collection?
 Each DNA collection should be split initially into 2 or 3
portions. One sample (plant material or extracted DNA) should
be stored at a working (dispensing) node and another por-
tion(s) be stored in at least 1 (one), but desirably 2 (two)
back-up reserve (base) nodes. The reserve nodes should be in
different countries and if possible on different continents to
safeguard the DNA samples against various natural and man-made
catastrophes.

General Recommendations:
 a. DNA should be extracted from preserved materials only

when the DNA is needed. Delaying the extraction has the
advantage of letting technology catch up so advanced techniques can be used as they become available.

   b. Working nodes should generally be an existing organization with adequate biochemical expertise and have an associated herbarium. Having an herbarium on site would not be
required but a very close, local (in the city) association
with a recognized (see Index Herbariorum, ed. 8) is required.

   c. For working as well as reserve nodes, it is necessary
to have a strong institutional commitment, not just a personal
commitment, in order that the collection be maintained in
perpetuity not just for the lifetime of one person who has
committed himself to the idea.

   d. Consideration should be made concerning the availability of dependable electricity and liquid nitrogen in determining the feasibility of establishing a node.

   e. At a future meeting, it will be critical to discuss
the importance of local political stability in establishing
working and reserve nodes.

   f. Considerable interest was shown in the concept of
storing composite DNA samples (e.g., a composite of DNA from
all the legumes in a region, to be used for screening or
retrieval of unusual genes). No consensus was reached but
this needs to be addressed at a future meeting.

   g. The need for computer and data base compatibility was
expressed. Given the number of flat file and relational data
bases that are compatible with dBASE, it would seem that dBASE
compatibility would be desirable. No consensus was reached in
regards to this nor on the use of a flat file vs. relational
data base. It was felt that the critical issue at present was
to begin collecting DNA rich materials. Computer and data
base questions should be addressed at a future meeting.

GROUP B: SCOPE OF PLANT COLLECTIONS

Group members: Dennis Stevenson - Rapporteur, David Giannasi, Toby Hodgkin, Lin Zhong-Ping and Victor Villalobos,

Task 1. Can we prioritize plant species to collect? If so, what should have the highest priority?

Prioritizing Plant Species to Collect

Although all species are worthy of conservation, there is a need for an initial focus rather than random collections. Given this constraint, economically useful plants should be given some priority. However, this priority does not include the major crop plants of commercial usage that are widely cultivated (e.g. maize, rice, wheat, etc.) but rather those indigenous species that are tended and/or otherwise used by local people. The tended species include those used for indigenous agroforestry systems (e.g. amaranths), tropical timbers, fruit and nut trees. The tended species include plants utilized for medicinals, dyes, soap, fiber, spices, flavors, waxes, etc.

Prioritizing Geographical Areas to Collect

Geographical areas should be used as a guideline, following the IUCN system (see Heywood, this book). Collectors should consider the following factors and a considerable effort should be focused on the tropical rain forests and tropical highland regions. Among the factors to be considered in selection a geographic region to focus efforts on are: diversity of taxa; uniqueness of habitat; amount of depletion of taxa and genetic diversity that has already occurred; and areas in which institutions are **not** working to preserve diversity.

Task 2. What are some collection protocols that should be followed?

DNA collectors should be considered the same as all other plant collectors. Consequently they should:

    a. Voucher all collections in recognized herbaria (i.e., listed in Index Herbariorum, ed. 8)

b. Provide proper label information as to the locality, habitat, etc. for each plant collected.

c. Follow all procedures concerning permits, convenios, and deposition of duplicate vouchers in the country of origin.

d. Collect leaf samples and pack them in desiccants (see Adams, Nhan and Chu, this book) immediately (the same day). Leaves are of value as simple long-term storage. In addition, DNA may be retrieved from existing herbarium collections as techniques are refined. This resource should not be overlooked.

e. In the case of legumes, samples of root nodules should be taken if possible, but kept as a separate accession.

f. If a chemical treatment is used in the field, information should be provided concerning the method and some untreated leaves must be stored in desiccant (see d. above).

Task 3. What should be the taxon scope of collections?

Collections should initially be representative at the generic level with priority given to monogeneric families. Specifically:

a. Fungi - Taxa should be first checked with Type Culture collection to determine if they are documented and representative (e.g. American Type Culture Collection).

b. Algae - See fungi (a. above)

c. Bryophytes - Consideration should be given to published lists concerning the area in which collecting is being done.

d. Vascular Plants - In the case of phylogenetic relicts etc., material from garden origin will suffice. In general they will represent monogeneric families (e.g. *Ginkgo*, *Austrobaileya*). Otherwise, the large scale floristic surveys being conducted by institutions such as the Missouri Botanical Garden, New York Botanic Garden, Royal Botanic Garden - Kew, and others, would seem to provide an ideal opportunity for the collection of vast numbers of tropical species without the need for large additional funds.

e. Fossil plants - When possible, fossil material should be included. In this case, when destruction of the source material occurs, documentation via photographs and fragments is necessary.

All of the above are intended as guidelines and for the initial establishment of DNA Bank-Net.

RECOMMENDATIONS FOR RESEARCH

There is a need for research on the use of various types of sampling. For example, perhaps the most simple method would be to use carefully dried leaf material that is then cryo-preserved.

The vast resources of herbarium, dried specimens may hold considerable DNA that would be suitable for PCR. Because there are many types of herbaria storage, preservation and collections, there is a need for systematic investigations of the effect of modes of preparation, collection and storage on the integrity of DNA in the world's major holdings.

GROUP C: QUANTITY AND TYPES OF MATERIAL TO PRESERVE:
Group Members: Steve Price - Rapporteur, Tony Cox, Abebe
Demissie, James Miller and Peter Strelchenko

Task 1. How many genotypes per species should be stored?
The group recommended that as many genotypes as possible
be collected. Operationally, there will be overlap between
collectors. The consensus was that having one sample for each
of 1000 species would be much more valuable than having 1000
samples of one species.

Task 2. How many replicates should be stored per genotype?
For species, such as trees, an accession may be one individual. Yet for some species, such as duck weed, an accession
may be many individuals. In the case of plants that produce
more than 10 g (fresh weight) of leaves, three (3) samples
(replicates) should be collected and stored at three (3) different nodes. Of course, the amount of plant material that is
needed will be influenced by the efficiency of DNA extraction.

Task 3. What kinds of plant material should be preserved?
Operationally, leaves will probably be collected, but if
seeds are available, they should be stored also. Succulent
tissues or any plant part with fungal or bacterial (alien DNA)
contamination should be avoided. Strive for "clean" collecting.

**Task 4. Should we store plant organs, ground leaves, and/or
only DNA?**
The group recommended that at least the following should
be stored for each accession:
a.   Herbarium specimens (2 or more)
b.   Tissue/organs awaiting extraction and/or put into
reserve (base) nodes.
c.   Extracted DNA

GROUP D: PROJECTED USES OF DNA

Group Members: David Giannasi - Rapporteur, Tony Cox, Toby Hodgkin, Zheng Sijun and Lin Zhong Ping

Task 1. What are the current and projected uses of DNA from DNA Bank-Net?

The task group sees the following immediate uses for conserved DNA: Molecular phylogenetics and systematics of extant and extinct taxa; Production of previously characterized secondary compounds in trans-genic cell cultures; Production of trans-genic plants using genes from gene families; *In vitro* expression and study of enzyme structure and function; and Genomic probes for research laboratories.

Task 2. Logistical and Administrative problems.

The use of DNA depends on its supply and availability to the user. This depends on the logistics and administration of the network. The task group recommended that:

a. Each participating institution decide approximately how many species it will supply or be responsible for supplying within a geographic region (e.g. Will Kew be responsible for the "European flora", its vast collections from Africa and Asia, or just its own holdings in the Garden?)

b. The basic DNA or DNA rich stock should be divided into two (or more) portions for storage at a reserve node(s) as well as for use at the working (dispersing) node. The amplified, lyophilized DNA should be distributed in sealed ampules.

c. In the event that all the DNA stored at a working node has been dispensed or lost, the local participating member should make every effort to renew the basic stock from locally grown or natural plants.

d. A list of minimal charges for preparation, storage and shipping as lyophilized sample needs to be developed. Payments could be made in cash or exchangeable commodities (e.g. reciprocal DNA samples, books, or other exchangeable materials).

e. There needs to be developed a published set of extraction procedures for all participating Bank-Net members. A copy of the extraction protocol (or literature reference) should be sent with each DNA sample.

   f. Research is needed to develop a GAWTS (Genomic Amplification with Transcript Sequencing, Sommer, *et al.*, 1990)
type protocol (Fig. 2) for eventual supplementation of DNA
reserve stocks and obviate the need for replenishment from
outside sources.

High molecular weight genomic DNA

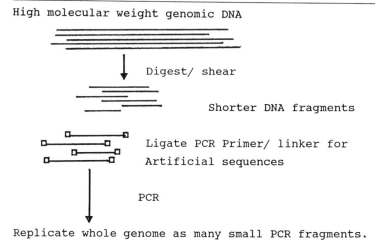

Figure 2. Diagram of the procedure for Genomic Amplification
With Transcript Sequencing (GAWTS) based in part on Sommer, *et
al.*, (1990).

---

   g. Similar approaches can be used for preservation and
distribution of DNA recovered from fossil plant specimens
except that the amount of DNA is very limited. Renewal of the
base stock after distribution of shipping stock can be very
difficult or impossible. Accessibility to regular distribution of fossil sources of DNA will become more feasible when a
GAWTS system is in use allowing for continuous amplification
from a single fossil DNA source.

   h. Research is needed to develop and extend methods for
amplification of immobilized genomic DNA by PCR. Figure 3
depicts a possible scheme for dispensing DNA copies without
losing the original DNA. This technique would allow multiple

DNA Immobilization

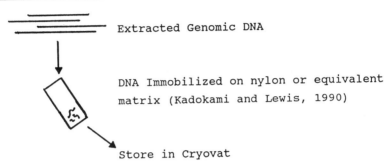

Extracted Genomic DNA

DNA Immobilized on nylon or equivalent
matrix (Kadokami and Lewis, 1990)

Store in Cryovat

DNA Amplification

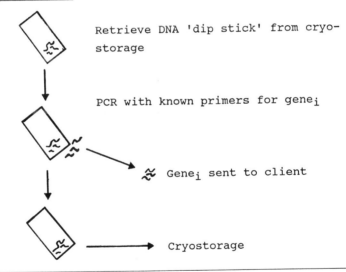

Retrieve DNA 'dip stick' from cryostorage

PCR with known primers for gene$_i$

Gene$_i$ sent to client

Cryostorage

Figure 3. Scheme for DNA immobilization on a 'dip stick' and
subsequent amplification (see Kadokami and Lewis, 1990).

probings and amplifications.

   i. Some material may be accessed from herbarium specimens
under control of local curators using current methods of DNA
extraction. Herbarium sheets should be marked if sampled for

DNA. Herbarium specimens are limited in supply and their
utility appears to be limited to material collected without
chemical preservation. Material may be sampled directly from
the sheet or the attached specimen envelope if it contains
sufficient leaf material (ca. 0.1 - 0.5 g dry wt.) for DNA
extraction. It seems likely that the integrity of DNA would
decrease with the age of specimens. Additional research is
needed on DNA stability in specimens stored for different
times and under different regimes in different countries.

Literature Cited

Kadokami, Y. and R. V. Lewis. 1990. Membrane bound PCR.
    Nucleic Acid Res. 18: 3082.
Sommer, S. S., G. Sarkar, D. D. Koeberl, C. D. K. Bottema,
    J-M. Buerstedde, D. B. Schowalter and J. D. Cassady. 1990.
    Direct sequencing with the aid of phage promoters. *In*: M.
    A. Innis, D. H. Gelfand, J. J. Sninsky and T. J. White
    (eds.), *PCR Protocols*. Academic Press, San Diego.

GROUP E. INFRASTRUCTURE NEEDS
    Group Members: Victor Villalobos - Rapporteur, Abebe
Demissie, Bart Panis, Phillip Stanwood and Dennis Stevenson.

The task group developed the following concepts to address
infrastructure needs. It was perceived that a series of
phases be developed to accomplish the objectives of DNA preservation and future utilization.

Task 1. Establishment of DNA Bank-Net Association
    There is a need for a professional organization that would
function initially as a lead organization and superstructure.
The association would bring together capabilities, expertise,
and financial support to produce a newsletter, publications,
coordination of DNA banking activities and develop sources of
funds for regional and national nodes in the network.
    A major objective in this phase would be the development
of global DNA Banking strategy and a series of specific objectives and action plans to achieve the overall strategy.

Task 2. Equipping and Training for the Nodes in the Network
    a. Funds need to be obtained to produce specific action
plans for DNA Bank-Net to enhance the funding opportunities
from groups such as foundations and international concerns.
    b. Technical workshops need to be conducted in order to
bring experts together and develop specific techniques and
protocols for DNA extraction, amplification and storage.

GROUP G: NATIONAL AND INTERNATIONAL RELATIONS

Group Members: L. A. Barreto de Castro - Rapporteur,
Daniel Abbiw, James Miller, Steve Price and Peter
Strelchenko.

Task 1. Transportation of plant materials in and out of countries.

The task group recommends that:

a. Regulations with respect to biological safety must be
followed.

b. Members are to respect a code of ethics as to guarantee that collections are only made with the knowledge and consent of the country.

c. DNA Bank-Net must be aware that countries have quarantine regulations in effect and DNA Bank-Net must not conflict with these regulations during its activities. This includes the requirements for the collection as well as the exportation of plant material.

d. Collaborations should be offered the opportunity for royalties and/or return profits/products or processes that may result from use of the plant materials collected--patented or not.

e. DNA Bank-Net should support the concept of using license agreements and/or contracts that insure that there is a flow of money back to the countries and/or institutions when commercialization of protected germplasm is achieved. This will guarantee the uninhibited exchange of germplasm for scientific purposes, so that scientific achievements are not obstructed.

Task 2. Interfacing with seed banks, botanic gardens and other related institutions.

DNA Bank-Net should complement activities already being performed by different institutions, specifically, those working in the area of germplasm collection and conservation. The interface with these institutions is necessary however for two additional reasons:

a. To generate incentives which will eventually attract institutions to operate in association with the DNA Bank-net.

b.   To identify institutions with potential to operate as either a working or reserve node (see definition in the Resource Identification Group report).

Task 3. Ownership and Proprietary Interests.

[Eds. note: The task group strongly supported the idea that DNA Bank-Net should become a non-profit international corporation. However no consensus was reached by the entire participants, as a considerable number of representatives favored a very loose association with more of an academic focus. The ideas of the task group are challenging, so we have included their report, essentially verbatim for future discussions.]

The task group envisions DNA Bank-Net as a public non-profit international germplasm conservation organization. Funding for the establishment and operation of DNA Bank-Net, Inc. could be obtained from:

a.   International funds that exist to support the conservation of plant diversity, such as IBPGR and other international funds for Biological Diversity Protection. These funds would include financial support for *in situ* germplasm conservation as well as the activities of the DNA Bank-Net. The participation of private funds is mandatory.

b.   Royalties which may result from licensing of genes could contribute an incentive for public institutions such as Kew, Missouri and NY Botanical Gardens to associate with the DNA Bank-Net. Royalties should be divided among participants, parties in benefit, and provide additional budgetary funds. Intellectual proprietary rights must be considered both to justify investments from private institutions as well as an incentive for developing countries to participate in the establishment of an International Fund for Biological Diversity Protection.

General Recommendations:

a.   The UNCED (United Nations Conference on Environmental Development) which will held in Rio de Janeiro in 1992 should be an interesting opportunity to present these ideas, world-wide, and to generate funds for the establishment of the DNA Bank-Net.

b. DNA banks are important not only for phylogenetical
reasons but also as a source of genes, probes, and promoters
which are both of scientific and commercial importance. DNA
collections of plant species in bulk (composite samples) may
be considered as an additional adjunct system that could provide genes from "horizontal" gene pools. That is, as gene
function is determined in the future, it will likely become
practical to select specific genes that have been identified
by gene tagging and other techniques and remove copies of
these from a composite of the DNA from numerous species.

The following individuals/institutions have expressed an
interest in DNA Bank-Net:

Dr. Robert P. Adams, Plant Biotechnology Center, BU Box 97372
Baylor University, Waco, TX 76798 USA
Phone 817-755-1159, FAX 817-752-5332

Dr. Daniel K. Abbiw, Botany Department, Box 55, University of
Ghana, Legon, Ghana, West Africa

Drs. Lucia Atehortua/Ricardo Callejas, HUA, Department of
Biology, University of Antiquia, Medellin, Colombia
Phone 57-4 263-0011 FAX 57-4-263 8282

Dr. Luiz Antonio Barreto de Castro, CENARGEN/EMBRAPA, Parque
Rural, CP 102372, Ave. W5 Norte, W 70770, Brasilia DF, Brazil
Phone 55-061-272-4203   FAX 55-061-274-3212

Dr. Mike Bennett, Director, Jodrell Labs, Royal Botanic
Garden, KEW, Richmond, Surrey TW9 3AB, England
Phone 081-940 1171 FAX 081-948 1197

Dr. Anthony H.D. Brown, Division of Plant Industry, CSIRO
GPO Box 1600, Canberra, ACT 2601, Australia

Prof. Cheng Xiongqying, Institute of Nuclear Agricultural
Sciences, Zhejiang Agricultural University
Hangzhou, 310029 China

Prof. Chu Ge-lin, Institute of Botany, Northwest Normal
College, Lanzhou, Gansu, China

Dr. Robert E. Cook, Arnold Arboretum, Harvard University
125 Arborway, Jamaica Plain, Mass. 02130-3519
Phone (617) 524-1718

Dr. Tony Cox, Jodrell Labs, Royal Botanic Garden, KEW,
Richmond, Surrey TW9 3AB, England
Phone 081-940 1171   FAX 081-948 1197

Dr. Marshall Crosby, Asst. Director, Missouri Botanical Garden
2315 Tower Grove Ave., St. Louis, MO 63166
Phone (314) 577-5169   FAX (314) 577-9596

Abebe Demissie, Plant Genetic Resources Center, P.O. Box
30726, Addis Ababa, Ethiopia
Phone 18-03-81   FAX 251-1-552-514
                         c/o Lutheran World Federation

Dr. Thomas Elias, Director, Rancho Santa Ana Botanic Garden,
1500 North College Ave., Claremont, CA 91711
Phone (714) 625-8767   FAX (714) 626-7670

Prof. J. Eloff, Director, National Botanical Institute,
Private Bag X101, Pretoria, 0001 South Africa
Phone (012) 86 1164   FAX 012 86-1194

Dr. Z. O. Gbile, Director, Forestry Research Institute of
Nigeria, Private Mail Bag 5054, Ibaden, Nigeria

Dr. David Giannasi, Dept. of Botany, University of Georgia,
Athens, GA 30602
Phone 404-542-1819; FAX 404-542-1805

Dr. David Given, Department of Science and Industrial Research
Botany Division, Private Bag, Christchurch, New Zealand
Phone 03 252 511   FAX 03 252 074

Dr. Chaia C. Heyn, Dept. of Botany, The Hebrew University,
Jerusalem, 91904, Israel

Dr. Vernon Heywood, IUCN Conservation Monitoring Center
53 The Green, Kew, Richmond, Surrey TW9 3AA, England
Phone 081 940 4547   FAX 018 948 4363

Dr. Toby Hodgkin, Research Officer, Genetic Diversity
IBPGR, c/o FAO of the UN, Via delle Sette Chiese 142
00145 Rome, Italy
Phone (39 6) 574 4719 FAX (39 6) 575 0300

Prof. Hu Zhong, Professor of Plant Biochemistry, Kunming
Institute of Botany, The Academy of Sciences of China
Heilongtan, Kunming, Yunnan, China
Phone 24053

Dr. Kunio Iwatsuki, Botanical Gardens, Faculty of Science
University of Tokyo, 3-7-1 Hakusan, Bunkyo-ku
Tokyo 112, Japan
Phone   FAX 81-33-814-0139

Prof. Mupinganayi Kadiakuida, Director General, CARI
B.P. 16513, Kinshasa, Republic of Zaire

Dr. S. L. Kapoor, Cytogenetics Lab, National Botanical
Research Institute, Rana Pratap Marg., Lucknow 226 001, India

Prof. Lin Zhong-ping, Division of Plant Molecular Biology
Institute of Botany, Academia Sinica, Beijing 100044, China
Phone 893 831 345   FAX 0086-01-8312 840

Dr. Ma Cheng, Chief Engineer, Bureau of Bio-Sciences and
Biotechnology, Chinese Academy of Sciences (Academia Sinica)
52 San Li He Road, Beijing, China

Dr. Lydia Makhubu, University of Swaziland, Kwaluseni Campus
P/Bag, Kwaluseni, Swaziland

Dr. John S. Mattick, The Gene Library, Centre for Molecular
Biology and Biotechnology, University of Queensland
St. Lucia, Queensland, QLD 4072 Australia
Phone (07) 365 4447 FAX 617 371-7588 or 365-4388

Dr. James Miller, Missouri Botanical Garden
2315 Tower Grove Ave., St. Louis, MO 63166
Phone (314) 577-5169   FAX (314) 577-9596

Dr. Titus K. Mukiama, Department of Botany, University of
Nairobi, Chiromo, P.O. Box 30197, Nairobi, Kenya
Phone 43181

Dr. F. Ng, Deputy Director, Forestry Research Institute
Malaysia (FRIM), Karung Berkunci 201, JLN FRI Kepong, 52109
Kuala Lumpur, Malayasia

Dr. Simon Owens, Jodrell Labs, Royal Botanic Garden, KEW,
Richmond, Surrey TW9 3AB, England
Phone 081-940 1171   FAX 081-948 1197

Dr. Mohinder Pal, Head, Cytogenetics Lab, National Botanical
Research Institute, Rana Pratap Marg., Lucknow 226 001, India

Dr. Bart Panis, Laboratory of Tropical Crop Husbandry,
Catholic University of Leuven, Kardinaal Mercierlaan 92.
B-3001 Heverlee, Belgium
Phone: 016-22-09-31; FAX: 016-22-18-55

Dr. Ghillean T. Prance, Director, Royal Botanic Garden, KEW,
Richmond, Surrey TW9 3AB, England
Phone 081-940 1171   FAX 081-948 1197

Dr. Steve Price, Industrial Liaison Officer, Office of Intel-
luctual Property, Iowa State University, Ames, Iowa, 50010
Phone: 515-294-4741; FAX: 515-294-0778

Dr. Mukunda Ranjit, Assistant Pomologist, Fruit Development
Division, Kirtipur, Kathmandu, Nepal

Dr. W. Roca, Head, Bio-Tech Unit, CIAT, AA6713, Cali, Colombia

Dr. Phillip Stanwood, National Seed Storage Lab, USDA,
Colorado State University, Ft. Collins, CO 80523
Phone (303) 484-0402   FAX (303) 221-1427

Dr. Dennis Stevenson, The New York Botanical Garden, Bronx,
New York 10458-5126
Phone (212) 220-8999   FAX (212) 220 6504

Dr. Peter Strelchenko, N. I. Vavilov Institute of Plant
Industry, 42, Herzen Street, 190000, Leningrad, USSR
Phone 314-48-48   FAX 812 311-8762

Prof. Thavorn Vajrabhaya, Dept. of Botany, Faculty of Science,
Chulalongkorn University, Bangkok 10500, Thailand

Dr. Victor M. Villalobos, Director, Programa Mejoramiento
Cultivos Tropicales, CATIE, Turrialba 7170, Costa Rica
Phone 506 56-0232   FAX 506 56-0606

Prof. Luz Ma. Villarreal de Puga, Instituto de Botanica,
University de Guadalajara, Apartado 139, Las Agujas, Nextipac,
Zapopan Jalisco, Mexico

Dr. Melaku Worede, Plant Genetic Resources Center, P.O. Box
30726, Addis Ababa, Ethiopia
Phone 18-03-81   FAX 251-1-552-514
                  c/o Lutheran World Federation

Prof. Zheng Sijun, Dept. of Agronomy, Zhejiang Agricultural
University, Hangzhou, 310029, Zhejiang, China
Phone 741 733-2240   FAX: 0086 571 774636